Chilton's
CB HANDBOOK

GUIDE TO CHOOSING, INSTALLING & OPERATING CITIZENS BAND RADIO

CARS • TRUCKS • RV's • MARINE • VANS

Prepared by the Automotive Editorial Department

president and chief executive officer **WILLIAM A. BAR-BOUR;** executive vice president **K. ROBERT BRINK;** vice president and general manager **WILLIAM D. BYRNE;** editor-in-chief **JOHN D. KELLY;** managing editor **JOHN H. WEISE, S.A.E.;** assistant managing editor **PETER J. MEYER;** senior editor **KERRY A. FREEMAN**

CHILTON BOOK COMPANY   Radnor, Pennsylvania

Copyright © 1976 by Chilton Book Company
First Edition
All Rights Reserved
Published in Radnor, Pa. by Chilton Book Company
and simultaneously in Ontario, Canada
by Thomas Nelson & Sons, Ltd.

Manufactured in the United States of America

Library of Congress Cataloging in Publication Data

Chilton Book Company. Automotive Editorial Dept.
  Chilton's CB Handbook.
  1. Citizens radio service. I. Title. II. Title: CB handbook.
TK6570.C5C46 1976   621.3845'4   76-10769
ISBN 0-8019-6512-8 pbk.
ISBN 0-8019-6513-6

Photography by Kerry A. Freeman

CONTENTS

ACKNOWLEDGMENTS

Chilton Book Company wishes to express appreciation to the following firms for their generous assistance in the preparation of this book.

The Antenna Specialists Company
Cleveland, Ohio 44106

Beltek Corporation
Gardenia, California 90247

Breaker Corporation
Arlington, Texas 70611

Browning Laboratories
Laconia, New Hampshire 03246

Champion Spark Plug Company
Toledo, Ohio 43601

Craig Corporation
Compton, California 90220

Dynascan Corporation (Cobra)
Chicago, Illinois 60613

E.F. Johnson Company
Waseca, Minnesota 56093

Fanon-Courier Corporation (Courier)
Perth Amboy, New Jersey 08861

G. C. Electronics
Rockford, Illinois 61101

Hy-Gain Electronics Corporation
Lincoln, Nebraska 68505

J.I.L. Corporation of America, Inc.
Carson, California 90746

Kris, Inc.
Cedarburg, Wisconsin 53012

Linear Systems, Inc. (SBE)
Watsonville, California 95076

Midland International
Communications Division
North Kansas City, Missouri 64116

New Tronics Corporation (Hustler)
Brook Park, Ohio 44142

Parsons Electronics and Vehicular Systems
Belvidere, New Jersey 07823

Pierce-Simpson
Division of Gladding Industries
Miami, Florida 33152

Rogers Electro-Matics, Inc.
Syracuse, Indiana 46567

Royce Electronics Corporation
North Kansas City, Missouri 64116

Shakespeare Industrial
Antenna Division
Columbia, South Carolina 29202

Sparkomatic Corporation
Milford, Pennsylvania 18337

Sprague Products Company
North Adams, Massachusetts 01247

Surveyor Manufacturing Corporation
Madison Heights, Michigan 48071

Telex Communications, Inc.
Minneapolis, Minnesota 55240

Tenna Corporation
Cleveland, Ohio 44128

Turner Company
Division of Conrac Corporation
Cedar Rapids, Iowa 52402

INTRODUCTION
TO CB RADIO

Few events have affected the life of the average motorist more than the Great Fuel Crisis of 1973-74 and the subsequent lowering of the national speed limit to 55 miles per hour. Ironically, it was the nationwide truckers strike, a direct result of the fuel crisis and strict enforcement of an unpopular 55 mph speed limit on interstate and intrastate highways, that spawned the current boom in Citizen's Band radios. In two years, they have become so popular that they now rank second only to the telephone as the largest form of two-way communication.

Truckers have been using Citizen's Band radios for years and part of the reason for their popularity is the national attention focused on the truckers' use of CB radios to keep in touch with each other during the truckers strike of 1974. The motoring public suddenly became aware of the complicated network of CBs keeping track of "Smokey Bear," among other things. Thousands of motorists rushed out and bought CB radios to monitor truckers' conversations, receiving up-to-the-minute information on traffic and weather conditions and speed traps.

Since then, CB radio has become almost a subculture. CBers consider themselves an independent and rebellious group and strongly indentify with the long-haul truck drivers. The language and speech patterns of the truckers have become the "official" CB language on the road. If you don't talk the language, you might not find out about that speed trap over the next hill because no one will answer you. Legally, you're required to identify yourself by your station call sign. But, in the real world on the interstates, call signs go largely unheard. The "handle," or identifying nickname, is used instead and it is considered extremely bad form and the mark of an amateur to mention names on the air.

Many drivers who now own CBs wouldn't dream of going very

far without their radios. Aside from the obvious benefits of knowing the location of almost every speed trap for miles, it helps pass the time on a long, boring trip, and many drivers claim it actually helps keep them more alert.

In addition to the complex network of motorists and truckers keeping track of state and local highway patrol officers, CB radios serve hundreds of other invaluable (and not so invaluable) uses in cars, trucks, RVs, boats, motorcycles, and as base sets in the home or office.

• Campground operators frequently monitor Channel 11, anticipating campers searching for accommodations.

• Although CB is no substitute for marine VHF-FM radiotelephone, there are probably more boats with CBs than with marine radio. It's cheaper, better than nothing at all, and can keep you in touch with your home on shore or get you the local fishing hotspots for the day (a definite ''no-no'' on marine radiotelephone). Except for tidewaters, the Great Lakes, the Intracoastal Waterway and the St. Lawrence Seaway, there are no VHF or Coast Guard stations to communicate with, but you are surrounded by CBers on wheels and afloat.

• Garages, service stations, and private citizens monitor Channel 9 to assist in emergency situations or help stranded motorists. Channel 9 is the official emergency channel nationwide, which also includes such information as where to find food and lodging when traveling. Thousands of CB operators have banded together into organized groups and voluntarily monitor Channel 9 24 hours a day for the sole purpose of lending aid in emergencies. Among these are ALERT (Affiliated League of Emergency Radio Teams), HELP (Highway Emergency Locating Plan), REST (Radio Emergency Service Teams), and the largest, REACT (Radio Emergency Associated Citizens Teams), composed of over 1,000 teams and 40,000 members handling millions of emergency assistance calls annually. Members of these organizations have rendered invaluable service in almost all national disasters: earthquakes, floods, fires, tornados, and hurricanes. They are equipped to provide you with information or to route your call to the proper authority.

• Increasingly, state police are beginning to employ CB radio. Highway patrols in Illinois, Missouri, Georgia, and Ohio have or are contemplating installing CBs in patrol cars and some troopers buy their own sets. Police and CBers have been known to cooperate to apprehend hit-and-run drivers or drunk drivers. Speeders should also be aware that some truckers feel that anyone going considerably faster than they are, is a menace, and should be reported to the nearest "bear with ears" immediately.

THE CITIZENS RADIO SERVICE

In the late 1940s, 2-way radio communication was mainly restricted to governmental agencies, police, and fire departments. But, in 1947, the Federal Communications Commission formed the Citizens Radio Service, to permit 2-way communications over short distances by private individuals or businesses.

Presently, there are 3 CB classes—Class A, C, and D—each serving a different purpose. Originally, there was a Class B, but it no longer exists.

CLASS A

Very little activity existed on Class A in the beginning, possibly because very little equipment was available and tough regulations governed that. Equipment was fairly expensive, limited mainly to business and industry. Several channels were available, but, as today, license applicants had to specify which channel was desired. FCC approval is also needed to change to another channel. Currently there are 16 channels available:

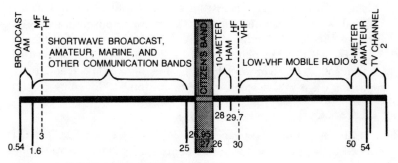

Location of the Citizens Band on the radio spectrum

CLASS A CB CHANNELS
(frequencies in MHz)

Frequency	Use
462.550	Base, Mobile
462.575	Base, Mobile
462.600	Base, Mobile
462.625	Base, Mobile
462.650	Base, Mobile
462.675	Base, Mobile
462.700	Base, Mobile
462.725	Base, Mobile
467.550	Mobile Only
467.575	Mobile Only
467.600	Mobile Only
467.625	Mobile Only
467.650	Mobile Only
467.675	Mobile Only
467.700	Mobile Only
467.725	Mobile Only

Transmitter input power is restricted to a maximum of 60 watts (48 watts output) on either AM or FM radiotelephone, which constitutes much of today's Class A CB equipment.

CLASS B

A Class B CB license permitted operation on 465 MHz only, and transmitter input power was restricted to a maximum of 5 watts. Class B units were rare until the 1950s when a few manufacturers began making reasonably priced sets. But difficulties arose. Sets were generally low on power, stemming from the fact that technical specifications were less stringent if transmitter power was less than 3 watts. The range of these low-powered uhf sets was extremely limited and impractical. It was found that a radio signal at 30 MHz will achieve better results under identical conditions than one at 450 MHz, so in 1968 the FCC abandoned Class B and forced existing Class B stations to cease operation by November, 1971.

CLASS C

Classes A and B were followed shortly by Class C. Class C is specifically intended for radio control of models, garage door openers, etc., and no voice communication is allowed. Transmitter input power is restricted to 5 watts, except on 27.255 MHz, where the limitation is 30 watts.

CLASS C CB CHANNELS
(frequencies in MHz)

Frequency	Use
26.995	1
27.045	1
27.095	1
27.145	1
27.195	1
27.255	1 Shared with stations in other services
72.08	3
72.16	2
72.24	3
72.32	2
72.40	3
72.96	2
75.64	3

Use 1: Radio control of remote devices or remote radio control of any device used to attract attention;

Use 2: Radio remote control of any model used for hobby purposes;

Use 3: Radio remote control of model aircraft only.

These frequencies are subject to the conditions that their use will not (1) cause interference with the remote control of industrial equipment operating on the same or adjacent frequencies; (2) cause interference with scientific, medical, or industrial devices operating in the 26.96-27.28 MHz band; and (3) will not affect the reception on television Channels 4 or 5.

CLASS D

CB radio, as it is known today, officially was created in 1958, when the Federal Communications Commission allotted 22 channels (frequencies), of the formerly 11 meter amateur broadcast band to provide reliable 2-way communication for private citizens. Class D stations operate exculsively on these frequencies, except for Channel 23, which was added later. Channel 23 is used on a shared basis with Class C stations.

The lower operating frequencies substantially overcame the line-of-sight transmission difficulties experienced with Class B equipment and the popularity of CB began its slow ascent as low-priced, reliable equipment was mass produced.

The FCC specifies radiotelephone only on the 23 Class D AM channels with maximum transmitter output power limited to 4 watts (12 watts peak envelope power for single sideband sets).

CLASS D CB CHANNELS
(frequencies in MHz)

Channel Number	Frequency
1	26.965
2	26.975
3	26.985
4	27.005
5	27.015
6	27.025
7	27.035
8	27.055
9*	27.065
10	27.075
11**	27.085
12	27.105
13	27.115
14	27.125
15	27.135
16	27.155
17	27.165
18	27.175
19	27.185
20	27.205
21	27.215
22	27.225
23***	27.255

* Channel 9 is the national emergency channel
** Channel 11 is the national calling channel
*** Channel 23 used on a shared basis with Class C stations

CB RADIO AND YOU

Every day, more and more people are turning to CB radio for pleasure and practical use. It has come a long way since the days when it was difficult to raise another station on the air. Really, though, all it means is that there are more people competing for the available air time. Legislation is being pondered in Washington to create more CB channels and possibly new equipment. But don't worry—most FCC rules and regulations don't take effect until 3-5 years after they have been legislated.

In the meantime, there are a few suggestions to make things easier on everyone. Some of these "rules" are written (see Part 95 of the "Federal Communications Commission Rules and Regulations"), some are unwritten and some are just common sense.

LICENSE

No operator's license is required for CB, as it is for ham radio. On the other hand, the FCC requires that you obtain a Citizens Radio Service Class D station permit before operating your rig. The only exceptions to the license requirement are those sets with transmitter power of 100 milliwatts or less. These are generally walkie-talkie type and require no license.

Generally, an FCC license application comes with each new CB set. Generally, as long as you are 18 or older and an American citizen, all you need do is fill out the application and send it in with your $4.00. If you are under 18, a parent or guardian can obtain a license and permit you to operate it. Once the station license is obtained, anyone can use the set, as long as the holder of the license gives his/her permission to do so.

Obviously, if you send in your application when you get your set, it is going to be a while before you can legally operate it. To avoid the hassle of waiting several months (that's about how long it takes these days), for a license and call sign, here's what to do.

First, if you are even dreaming about a CB set, write to the Federal Communications Commission, Washington, D.C., 20554, and ask for FCC Form 505. This is the CB Class D station license application. Also ask them to send along several FCC Form 452-C transmitter identification tags. You really don't have to have this exact form (a facsimile will do, See Appendix), but while you're about it, you might as well ask for it. It's self-adhesive and ready to use.

Second, write to the Superintendent of Documents, Government Printing Office, Washington, D.C., 20402, and ask for Volume VI of the Federal Communications Commission Rules and

Regulations. This contains Part 95 which governs CB Class D. It will cost $5.35, but it is necessary to obtain it; when you sign your license application, you affirm that you have, or have ordered, a current copy of the FCC Rules & Reg's., Part 95. When you purchase Part 95, you will also automatically receive future updates and rule changes.

Third, fill out the application form and send in your $4.00. You

---

United States of America
Federal Communications Commission

Form Approved
GAO No. B-180227(R01 02)

FCC FORM 505

August 1975

APPLICATION FOR CLASS C OR D STATION LICENSE IN THE CITIZENS RADIO SERVICE

INSTRUCTIONS

A. Print clearly in capital letters or use a typewriter. Put one letter or number per box. Skip a box where a space would normally appear.

B. Enclose appropriate fee with application. Make check or money order payable to Federal Communications Commission. DO NOT SEND CASH. No fee is required of governmental entities. For additional fee details see FCC Form 76-K. or Subpart G of Part 1 of the FCC Rules and Regulations, or you may call any FCC Field Office.

C. Mail application to Federal Communications Commission, P.O. Box 1010, Gettysburg, Pa. 17325

NOTICE TO INDIVIDUALS REQUIRED BY PRIVACY ACT OF 1974

Sections 301, 303 and 308 of the Communications Act of 1934 and any amendments thereto (licensing powers) authorize the FCC to request the information on this application. The purpose of the information is to determine your eligibility for a license. The information will be used by FCC staff to evaluate the application, to determine station location, to provide information for enforcement and rulemaking proceedings and to maintain a current inventory of licensees. No license can be granted unless all information requested is provided.

---

1. Complete **ONLY** if license is for an Individual or Individual Doing Business AS

FIRST NAME   INIT   LAST NAME

2. DATE OF BIRTH

MONTH   DAY   YEAR

3. Complete **ONLY** if license is for a business, an organization, or Individual Doing Business AS

NAME OF BUSINESS OR ORGANIZATION

4. Mailing Address

4A. NUMBER AND STREET

NOTE:
Do not operate until you have your own license. Use of any call sign not your own is prohibited

4B. CITY   **4C.** STATE   **4D.** ZIP CODE

5. If you gave a P.O. Box No., RFD No., or General Delivery in Item 4A, you must also answer items 5A, 5B, and 5C.

5A. NUMBER AND STREET WHERE YOU OR YOUR PRINCIPLE STATION CAN BE FOUND (If your location can not be described by number and street, give other description, such as, on RT. 2, 3 mi., north of York.)

5B. CITY   **5C.** STATE

6. Type of Applicant (Check Only One Box)

☐ Individual   ☐ Association   ☐ Corporation

☐ Business Partnership   ☐ Governmental Entity

☐ Sole Proprietor or Individual/Doing Business As

☐ Other (Specify) _____

7. This application is for

☐ New License

☐ Renewal

☐ Increase in Number of Transmitters

IMPORTANT
Give Official FCC Call Sign

8. This application is for (Check Only One Box)

☐ Class C Station License (NON-VOICE—REMOTE CONTROL OF MODELS)

☐ Class D Station License (VOICE)

9. Indicate number of transmitters applicant will operate during the five year license period (Check Only One Box)

☐ 1 to 5   ☐ 6 to 15   ☐ 16 or more (Specify No. and attach statement justifying need.)

10. Certification   I certify that:

• The applicant is not a foreign government or a representative thereof.

• The applicant has or has ordered a current copy of Part 95 of the Commission's rules governing the Citizens Radio Service. See reverse side for ordering information.

• The applicant will operate his transmitter in full compliance with the applicable law and current rules of the FCC and that his station will not be used for any purpose contrary to Federal, State, or local law or with greater power than authorized.

• The applicant waives any claim against the regulatory power of the United States relative to the use of a particular frequency for the use of the medium of transmission of radio waves because of any such previous use, whether licensed or unlicensed.

WILLFUL FALSE STATEMENTS MADE ON THIS FORM OR ATTACHMENTS ARE PUNISHABLE BY FINE AND IMPRISONMENT. U.S. CODE, TITLE 18, SECTION 1001.

11. _____
Signature of: Individual applicant, partner, or authorized person on behalf of a governmental entity, or an officer of a corporation or association

12. Date _____

FCC license application form

don't have to own a set to get a license, but your license is good for 5 years in the event you do obtain a set.

A word about the application form(s). There are many and they are all different. You may find any of them enclosed with a new set you might buy. The older one is about 6 pages long (with instructions and worksheet) and considerably more complicated than the newer versions. The newer ones are dated September 1974 or August 1975, are revised and simplified to encourage more people to apply for licenses, and are to be commended for their brevity. Any version of the three will get you a license, but the two newer ones are far simpler to use.

Be sure that you fill it out correctly, neatly and legibly. The FCC clerks will return it if you don't.

The form is basically self-explanatory, but the following directions may help in filling out the new version.

- PRINT IN CAPITAL LETTERS OR USE A TYPEWRITER.
- PUT ONLY ONE LETTER IN EACH BOX.

STEP 1: Print or typewrite your first name, middle initial, and last name, in that order.

STEP 2: Fill in your date of birth. If you were born November 26, 1945, it should appear—11 26 45.

STEP 3: Ignore this step if you are applying as an individual. If you are applying as a business, give the name of the business, skipping a box between words. You can print outside the boxes, if necessary, but be neat and legible, keeping the spacing approximately the same.

FCC transmitter identification card (Form 452-C)

STEP 4, (4A-D): This is your mailing address. If you are applying as a business, give the business mailing address. In Step 4, skip
a space between the number and street name.

STEP 5, (5A-C): This is the location of your transmitter records. You do not have to fill this in *unless* you gave a Post Office
Box number, RFD number, or General Delivery address in Steps
4-4D.

STEP 6: Check the appropriate box, but check only one box.

STEP 7: Check the appropriate box again. This part of the form
is also used for license renewal or to increase the number of
transmitters covered by your station license. If you want more
than 15 transmitters, you'll have to attach a written explanation of
why.

Also, if you are renewing an existing station license, give your
call sign.

STEP 8: Since you are reading this book, it is assumed that you
are interested in a Class D (Voice) Station license. Check the Class
D box.

STEP 9: Decide how many transmitters you want and check the
appropriate box. The average person should probably check 1-5.

STEP 10: Read what you are signing in Step 11.

STEP 11: Sign the application.

STEP 12: Don't forget to date the application.

Stuff the whole thing in an envelope with a check for $4.00 made
out to the Federal Communications Commission, and mail it to the
Federal Communications Commission, P.O. Box 1010, Gettysburg, Pennsylvania 17325. Do not mail it to Washington, D.C. You
will receive, by mail, your license and call sign, which is good for a
period of 5 years from date of issuance unless it is revoked for
cause. Don't worry if it takes a little time to get your license back.
The FCC is currently backlogged with about 250,000 license applications.

When you receive your license, you do not have to keep it with
the set, although, if you have a base station in your home or office,
the license will probably be where the set is. The license should be
kept at the "principal station" address given on the license application form (Item 4 or 5). If your set is installed as a "mobile" in a
car, truck, boat, or whatever, the set should have some kind of
identification—Form 452-C or a facsimile. This is a small card or
tag which lists the name of the license, where the license is located,

class of station, license call sign, signature of licensee and date of license expiration. This information is required to be on the form or facsimile and is normally sufficient proof of a license.

CHANNEL USAGE

All 23 Class D CB channels are now available for use between units of any station. Previously Channels 1-8 and 15-22 were reserved for communication between units of the same station and Channels 9-14 were reserved for communication between units of different stations. You can now use any channel with the following exceptions.

CHANNEL 9

In 1969, the FCC designated Channel 9 as the official emergency channel. The word "emergency" also applies to requesting route information and the availability of food or lodging as well as to more traditional "emergency" matters.

To avoid congestion on this channel, once contact has been established, communication should be moved to another channel. In the past, some organizations have assumed that because they are conducting "emergency" operations, blanket authority to use Channel 9 is given. This is not so. Emergency teams that have set up their operation in advance should use another channel and monitor Channel 9, if desired. Sets are also available that have a "Channel 9 priority" for this kind of operation, but more about that later.

CHANNEL 11

Effective September 15, 1975, Channel 11 is the official national calling channel. After initial contact has been established on Channel 11, communication should be moved to another agreed upon channel, except Channel 9. This is designed to reduce congestion on all of the other channels, but in practice, many CBers monitor another particular channel. This is because they cannot establish contact with anyone who might have the information they seek unless the other party is also on Channel 11. A good example of this is Channel 19.

CHANNEL 19

Channel 19 is the unofficial nationwide channel of the truckers, who adopted it as their own when it was discovered that they were bleeding over onto Channel 9 from their previous channel, 10. There is no official sanction to Channel 19, but all truckers know that if they want to contact another trucker, it can be done on Channel 19. Motorists also know that they can get faster results from Channel 19 than from Channel 11.

CHANNEL 13

In many areas of the country, Channel 13 is used by boaters as a marine calling channel. This is not officially sanctioned by the FCC, but it performs much the same function as Channel 11 on land. It also does not apply to all areas of the country, so it's wise to check beforehand with local residents.

CALL SIGNS & "HANDLES"

The FCC requires that you identify your station by the call sign given you when your station license was granted. The call sign consists of three letters followed by 4 numbers, and when it is given the FCC further requires that it be spelled out. For example, KST 6535 is "K-S-T-six-five-three-five," not "K-S-T-sixty-five-thirty-five."

Unit numbers should also be assigned to each transceiver, if you have more than one. Most times, your "base set" will be Unit 1. When you are communicating with another station (covered by another license), you need only use your call sign. When two units of the same station (covered by the same license), are communicating it should be:

"KST 6535 Unit 1 calling Unit 2."

"KST 6535 Unit 2 back to Unit 1."

Many CBers today identify themselves by colorful "handles" or nicknames, *without using a call sign*. The FCC definitely frowns on this and it is strictly contrary to the FCC Rules and Regulations. But as of September 15, 1975, the FCC has approved the use of "handles" *in addition to the station call sign*.

10-CODES

Most users of two-way radio are at least familiar with some of the ten-codes. The 10-codes are based on a system established by the

Associated Public Safety Communications Officials and are designed to conserve time on the air. Instead of, "Does anybody out there know what time it is?", "10-36?" will get the message across.

The revised APCO 10-codes contain only 34 codes, instead of the previously used 55 or so. Although there is no official Citizens Band 10-codes, the APCO 10-codes are about as much used as anything. Even so, there are local variations. For instance, many CBers use "10-36" for a time check, instead of the APCO "10-34," but time and use will familiarize yourself with most of the commonly used codes.

See Glossary for list of APCO 10-codes used by many CB'ers.

USING THE MICROPHONE

The microphone is a critical part of your set and for maximum results the mike must be used correctly.

Speak into the microphone holding it as a 45 degree angle, a couple of inches from your mouth. Don't speak directly into the mike, but hold it off to the side and speak past it. Voices spoken directly into a microphone at close range are received garbled and unintelligible. On the other hand, if the mike is too far away from your mouth, the modulation level will be too low and more background noise will be picked up, even though the S meter on the contact's receiver will indicate a fairly strong signal. When the modulation level is too low, your voice will sound weak and indistinct at the receiver. A little experimentation may be necessary at first to find the right spot for the best modulation but, a good rule of thumb to remember is: doubling the distance from your mouth to the mike decreases the mike's output by about 6 decibels or to about ¼ of its former output. This is one reason that the telephone handset and the headset were developed—to assure maximum modulation.

Speak in a normal voice, you don't have to shout. If you are using the mike properly, your modulation will be fine. Some sets are equipped with a microphone gain switch which can be adjusted for best modulation. Here again, experimentation with individual sets is necessary to get best results. The mike gain switch is covered under "selecting a set" in Chapter 2.

THE FCC AND THE LAW

The FCC is empowered by Congress to establish and enforce rules and regulations regarding the use of CB radios. Much of the problem with the present CB rules is that most people don't bother to read, or can understand, the 20-odd pages of legal jargon that constitute Part 95 of the FCC Rules and Regulations. However, in a move to simplify its rules and regulations, the FCC made the following modifications effective September 15, 1975.

 • The "hobby" restriction has been removed. The FCC will not issue citations for "chitchat," except in cases of profanity, playing music, or selling merchandise.
 • The use of "handles" is now approved provided the station call sign is also given.
 • All channels (except 9 and 11) may now be used for communication between any station.
 • Channel 11 is now the national calling channel.
 • Conversations are limited to no more than 5 minutes.
 • Maximum communication distance is 150 miles.

If you operate a CB set, there is really no excuse for not having a Class D station license, especially considering that it works out to about 80¢ a year at the current fee. But fear of Big Brother, or whatever reason, leads plenty of otherwise rational individuals to operate their CBs without a license. The situation is so bad that the FCC is considering issuing temporary licenses at the time of sale of the transceiver to cut down on the number of unlicensed transceivers.

The FCC is woefully understaffed to enforce its rules, maintaining a field staff over 400 people to keep track of millions of CBers. Not only must they monitor "base" stations, but they have to keep track of mobile stations also. It is a hopeless task, even though the state and local police frequently cooperate with the FCC mobile vans that patrol the highways with monitoring, recording, and direction finding equipment.

Even though the FCC establishes its own rules, it has no jurisdiction if you do not presently have, nor, have had, an FCC license. But, there is a big difference between violating an FCC rule and violating the "Communications Act of 1934, As Amended." An FCC inspector can cite you for a violation and refer the whole matter to the Justice Department for prosecution as a Federal offense. Admittedly, this seldom happens. The Justice Department has better things to do than prosecute individuals for operating a CB without a license, but occasionally it does happen; usually, in cases of blatant disregard for the rights of other licensees.

The FCC really gets upset about such things as failing to identify your station, operating a set without a license, obscenity on the air, and use of the "heater" or linear RF amplifier.

Obscenity on the air is fortunately a rare occurrence. If you're caught, penalties CAN run as high as $10,000 and a year in jail. Chances are they won't, but they could.

The linear RF amplifier is what the FCC really cracks down on. This is as device to boost the output of a CB from the legal maximum of 4 watts to 100 watts or more. These booster amplifiers have always been illegal for CBs, but the real trouble is that they blank out large geographic areas when in use, making it impossible for others to transmit. Their use is justified by some in emergency situations and they are expensive, running about "$1-2 a watt." However, the FCC does not sanction their use at any time, and on February 12, 1975, selling or owning a linear amplifier capable of being used on a CB radio became a federal offense. In fact, you can be fined for possession without even having the thing hooked up. If it is in your possession and it even looks like it could be used, you've committed a federal offense. To add insult to injury, the "heater" can be confiscated as evidence, your license can be revoked, and you can be fined. They get very upset about linears.

OPERATION IN FOREIGN COUNTRIES

Canadian CB regulations are similar to those in the U.S. Reciprocal agreements between the U.S. and Canada allow CB operation on both sides of the border. Canadians in the U.S. can obtain permission from the FCC and Americans traveling in Canada can contact the Canadian GRS (General Radio Service) for a tourist permit to use their transceivers while in Canada.

Mexico also allows CB operation, but like Canada, requires a permit.

DO'S AND DON'TS

The following should be observed by everyone:

1. DO NOT violate the sanctity of Channel 9. This channel is preserved strictly for emergency and motorist assistance calls.

2. DO NOT use Channel 11 for anything other than establishing contact. Once contact is established, move to another channel.

3. DO listen before transmitting. At least, ask for a break on a channel before transmitting, if you can't wait for a lull.

4. DO limit conversations to 5 minutes, except in emergencies. Otherwise, there are few conversations that require more than 5 minutes. Find out what you want to know and get off the air.

5. DO wait at least 1 minute before establishing a new contact.

6. DO observe the maximum communicable distance of 150 miles.

7. DO NOT exceed the maximum transmitter output power limitation of 4 watts on AM or 12 watts PEP on SSB.

8. DO NOT key the mike (push the mike button), unless you intend to use it (except when checking SWR). Keying the mike without talking into it produces an unmodulated or "dead" carrier and will block out the signals of nearby transmitter carriers which are modulated.

9. DO keep on-the-air testing to a minimum.

10. DO have your rig periodically checked to avoid transmitting off-frequency (bleeding over onto another channel).

BUYING A CB RADIO

There are literally hundreds of CB radios available on today's market. They range from 100 milliwatt walkie-talkies, with one channel only to AM/FM to CB converters all the way up to $1,000.00 base sets, equipped with clocks, meters, and other gadgetry. What kind of set you are interested in depends largely on what you are going to use it for and how much you want to pay. You can pay as much or as little as you like, but just buying an expensive set does not assure you of top performance. The purpose here is to acquaint you with most of the features available, to let you make up your own mind based on your needs.

CB's come in all price ranges from basic, uncomplicated mobiles to $1,000.00 and up base units. (Courtesy Browning Laboratories)

TYPES OF EQUIPMENT

Modern CB equipment, which is to say almost anything sold today, is loosely divided into 3 basic types: base, mobile, and portable.

17

Typical base set with all the goodies (Courtesy Royce Electronics Corp.)

Base stations are generally desk or table top transceivers with full solid state circuitry (no tubes). A few models still use some tubes which contributes to their higher consumption of power and larger physical dimensions. Because size is really no object with base sets, they generally offer slightly better performance than a mobile and are usually equipped with more luxury features. On the other hand they are generally more expensive (as are the antennas), than a mobile set. Almost all base sets are designed to operate from AC house current, but a few combination base/mobile transceivers are available to operate from 12 volt DC or 115 volt AC.

Mobile units are usually 23 channel transceivers of the type found in cars, trucks, boats, motorcycles, and RVs. These sets are totally solid state, since size is a limiting factor, as well as the low current drain on a car's battery made possible by solid state circuitry. Most mobile sets made today are designed to operate from 12 volt negative ground electrical systems, although, fortunately, some are capable of being switched from positive to negative

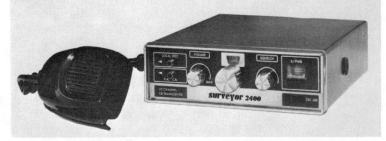

Typical 23-channel mobile transceiver (Courtesy Surveyor Manufacturing Corp.)

2-channel, 1 watt walkie/talkie (Courtesy Surveyor Manufacturing Corp.)

3-channel, walkie/talkie with full legal power (Courtesy Surveyor Manufacturing Corp.)

ground. Almost all cars, trucks, road bikes, RVs, and marine electrical systems made today are 12 volt negative ground, although there are some cars and trucks still on the road with either 6 or 12 volt, positive or negative ground.

Portable sets, or walkie-talkies, are considered to be capable of two-way communication while being carried by one person and capable of operation while in motion. Also known as hand-held transceivers, they are often used by children or adults where range is really no factor. They are rated from 100 milliwatts (which requires no station license), all the way to full 4 watt output sets, and are available with all 23 channels. A telescoping whip antenna is usually standard, as are batteries for a power source. Current drain on the batteries is low, except when transmitting, and in units which see a lot of service, rechargeable nickel cadmium batteries are a good investment.

TRANSCEIVER FEATURES

The word transceiver here is used synonymously with CB radio, although there are a few sets on the markets which are receivers only (they do not have transmit capability). These sets are rare these days, but if that's what you want, they are available. Some are even available to convert an AM or AM/FM radio into a CB receiver.

CB's are very competitive, available for all uses and budgets.

The idea here is not to sell you on certain features, but to explain what these features do, why they are important (or not important), and let you make up your own mind. Most CB radios look pretty much alike and are generally extremely competitive within a given

Equipment is available to convert AM/FM receivers into 23-channel CB receivers using the AM/FM antenna (Courtesy Tenna Corp.)

price range. The best advice is to read this and then go out to the local CB stores and pick up all the brochures you can and study them. Talk to the salesman and talk to people who already own CBs. Above all if you have a question try to find someone who is knowledgeable with CBs and who you trust and listen to his advice. You will probably find that to certain, opinionated questions,

Some manufacturers offer "package deals"—everything you need—transceiver, antenna and hardwaretesy Pace Communications, Div. of Pathcom, Inc.)

FCC type acceptance sticker

such as, "Who makes the best set?," "What kind of antenna should I get?," or "How much antenna cable should I use?," you will get a different answer from every person you ask.

Some manufacturers are now selling complete packages containing the transceiver and aerial to avoid shopping around for an aerial. If you go this route, be sure that the aerial is the type you want and/or can use. Be sure that any transceiver you buy has an FCC type acceptance sticker affixed.

Likewise, the argument rages as to where is the best place to buy a set. All electronic hobbyist type shops sell CBs, as well as most of the large chain stores. Some "old pros," who have been into CBs

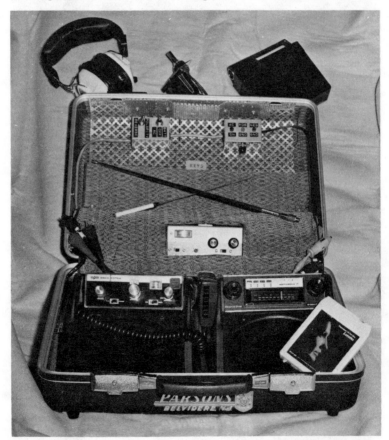

A completely portable, temporarily installed "suitcase" system, primarily for truckers. (Courtesy Parsons Electronics and Vehicular Systems)

for years, will warn you that a chain or department store does not have the experience in CBs that an electronics specialty store has. But, many department type stores are training their people in CB radio, so that they are fairly well versed and able to answer your questions. On the other hand, while electronics specialty stores are able to offer more personalized service (many one and two-man operations even offer installation in the price), their prices are sometimes slightly higher. As a customer you will want the best possible service, and electronics specialty stores probably have more experience with CB than anyone else.

CBs have become so popular of late that some of the car manufacturers have started offering CBs as dealer-installed options. General Motors and Chrysler have started programs and Ford and AMC are reported to follow soon. One company has even gone so far as to offer a completely portable system in a suitcase, catering to truckers constantly switching trucks. Clamp the antenna on the window and make the necessary connections with alligator clips, and you're ready to go.

RANGE

Normally, the first question someone asks about a CB set is, "How far can I get out (communicate)?." There is no single answer for this question, and anyone who tells you there is should be avoided.

CB radios are not powerful transmitters capable of reliable communication across the country. Attainable range is dependent on terrain, weather conditions, antenna placement on the vehicle (in the case of a mobile), how many people are using any given channel at a time, and type of equipment utilized to name a few factors. For example, range is much greater over the water where it is flat with no obstructions, than in a large city with tall buildings around. You will also notice that range increases greatly on a "clean" channel, one where few people are using their equipment. Placement of your aerial also has a lot to do with it, since most mobile CB antennas are directional, that is, they radiate the transmitted signal better in some directions than in others. See Chapter 3 for a further explanation.

As a general answer to the question, under optimum conditions, communication is reliable at 15-35 miles between base stations, at 10-20 miles between base and mobile stations, and 5-20 miles between mobile units. Sometimes, however, conditions will conspire against you, and you won't even be able to communicate around the corner.

There is also a phenomenon known as "skip" or "DX-ing." Basically, this involves a transmitted signal which will actually bounce off the inonosphere, once (single skip), or more than once (multiple skip), and finally be received at a transmitter several hundred or even thousands of miles distant. This does not always happen. Depending on the conditions in the ionosphere, some 50-400 miles above the earth's surface, these radio waves may simply be bent or reflected back to the earth's surface. Constant changes in the ionized layers of the earth's atmosphere vary the effects of "skip" considerably, and these changes may occur at predictable intervals (season-to-season, night-to-day), or it may be completely random.

OPERATING CONTROLS

The simplest transceivers have only 3 operating controls—a channel selector switch, a squelch control, and an on/off-volume control contained in one switch. The microphone has an additional control: a push-to-talk switch. As sets become higher-priced, you will find many more additional features, and before you buy a set, you should be aware of what these controls do, so that you can talk intelligently about them.

Basic 23-channel transceiver—meter, on/off volume switch and squelch (Courtesy Royce Electronics Corp.)

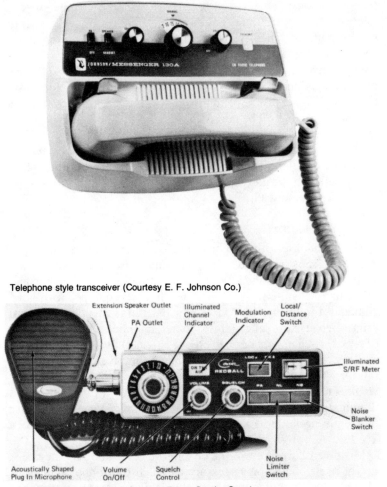

Telephone style transceiver (Courtesy E. F. Johnson Co.)

Popular transceiver features (Courtesy Fanon-Courier Corp.)

On/Off-Volume

This switch is usually combined into one control on the front of the set. Turning the switch clockwise turns the set on, just like any transistor radio.

Squelch

The squelch control is used to adjust the sensitivity of the receiver portion of the set. Normally, this is a single control, although on some sets it is the outer portion of the volume switch. Almost all sets made today have a squelch control. When it is set in the fully unsquelched position, you will hear a scratching static sound through the set which is being generated by the receiver itself. When a signal is received, it over-rides the noise and "turns off" the squelch circuit, allowing the signal to be heard. Normally, the squelch is set just beyond the point where background noise is silenced; nothing will be heard through the speaker until a signal is received. Rotating the squelch progressively toward the fully squelched position means that a stronger signal must be received to break the squelch. The converse is also true; the less the set is squelched, the weaker the signal it will receive, until the noise over-rides the squelch.

Channel Selector

The channel selector can be a rotating knob or a 3 position slide switch. A 3-position slide switch is normally used for sets equipped with only three channels, where a rotating knob is used for any more than 3 channels. You will notice that sets equipped with 3 or 6 channels, for example, are marked with letters instead of numbers. This is because the set generally comes equipped with only one set of crystals (it can only be used on one channel). In order to use it on another channel, you will have to install (or have installed), crystals for whatever channel you desire. Full 23 channel sets have the channel numbers marked on the knob, usually with a single light to illuminate the number of the channel in use. A few newer sets use light emitting diodes (LEDs) to show channel numbers, much like a TV set.

Most transceivers have the channel selector knob on the transceiver itself, making it necessary to reach down and change the channel. One manufacturer has solved this problem by putting the channel selector switch in the microphone housing, making one-handed control possible.

Automatic Gain Control (AGC)

Every receiver has an automatic gain control. The AGC is not really an operating control, but a feature of the circuitry to prevent overloading the receiver with an extremely strong signal. If the volume were turned all the way up to receive a particularly weak signal, you would be blasted out of the cab when a strong signal was received. The AGC circuit prevents this. The smaller the AGC specification, the better; 5 db, for example, is better than 10 db.

CB/PA Switch

Most CB sets come standard with the capability to be used as a public address system. If so, the set is equipped with a CB/PA selecting switch and a PA jack, usually on the back of the set. By connecting the jack to an external speaker, you can flick the switch to PA, talk into the mike, and use your set for a PA system.

An external PA speaker can be plugged into the set for public address use

External Speaker Jack

The external speaker jack serves the same function on a CB as it does on your home stereo. If you are not satisfied with the sound from the CB speaker built into the set, you can by-pass it by plugging an external speaker into the jack, usually on the back of the set.

Automatic Noise Limiter (ANL)

The automatic noise limiter is usually an on/off switch located on the front of the transceiver. Its function is to act as a filter, chopping holes in the received signal and substituting periods of silence, thereby reducing the static that the receiver picks up from man-made sources, such as car ignition, machinery, etc. Switching the ANL on will cut down on man-made static noise and will also sometimes cut down on received signal strength, however, only to a minimal degree. The automatic noise limiter is sometimes built-in, and should not be confused with the built-in automatic gain control, nor with the noise blanker.

Noise Blanker

The noise blanker is really a more powerful automatic noise limiter; and if it is switchable, is used, when the ANL does not adequately suppress man-made noise. It will also suppress the pulsing noise of your set.

Meter

If your set has a meter at all, it will probably be an S-meter, or, if a more expensive set, an S-RF meter. An S-meter indicates the relative strength of incoming signals in S-units, measured from 1-9 and in db above S9 when receiving. An S1 received signal would be very weak and an S9 received signal would be quite strong. Likewise, a signal metered at 40 db above S9 would be extremely strong.

The RF portion of the meter, if it has one, is a measure of the signal you are putting out. When this portion of the meter reads at maximum, it indicates that your rig is operating at peak efficiency.

The S/RF meter is an approximation only. Frequently you will hear someone ask for a radio check and a meter reading. This is OK for an approximation of how well your set is getting out, but modulation at the other receiver is a better indication. One S meter may read S7 for a received signal and another meter may read S5 for the same received signal.

Some meters also incorporate lights which glow in the meter to indicate various modes. For instance, a meter may glow red when you are transmitting or glow amber when you are receiving.

Some of the more expensive mobile transceivers and many base sets have a built-in SWR meter. This enables the operator to constantly monitor the condition of the antenna and spot defects in the system—more about this in Chapter 3. Check the accuracy of the SWR portion of the meter against an external VSWR bridge (meter).

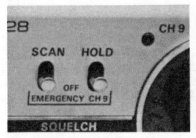 

Channel 9 priority controls (Courtesy Dyna-Scan Corp.)

Delta tune control to correct for off-signal reception (Courtesy DynaScan Corp.)

Channel 9 Priority

A Channel 9 priority is a device that enables the receiver to automatically override whatever channel is being utilized when a signal is received on Channel 9. Sometimes, a light will glow, alerting you to switch to Channel 9 or it may automatically do this when a signal is received. In this way, you can tune to any channel desired, and still monitor Channel 9. Normally, this feature is activated by means of an on/off switch.

Delta Tune

This control is a three-position switch on the front of the transceiver which allows you to correct your receiver for off-channel signals. Normally, it is left set at its center position, but when a signal is received with a high or low frequency error, the control is adjusted to either the plus (+) or minus (−) position. This allows you to receive off-frequency signals with a minimum of distortion.

DX/Local Switch

This is usually a two-position slide switch, enabling you to only receive local or short range calls if you're on the local side of the switch, or to enable you to receive longer range calls if you're on the DX (distance) side of the switch.

Mode Indicator Lights

Some transceivers are equipped with mode indicator lights; small lights, usually red and amber, which glow depending on whatever mode (transmit or receive) you happen to be in.

Tone Control

Just as most AM radios are equipped with a knob to adjust the tone of the sound (treble-bass), some CBs are also equipped with a tone control.

Microphone Gain Control

This is a knob that you will find on relatively few transceivers. Its use is to vary the percentage of modulation, thereby varying your talking power. Because the FCC limits the percentage of modulation to 100%, they are considering with holding type acceptance to receivers with a microphone gain control accessible to the operator of the set.

RF gain control to control modulation (Courtesy DynaScan Corp.)

Dynamic microphones usually plug into the transceiver

Microphones

The microphone is your connection between you and your transceiver. As such, it should be a good one. Most stock mikes that come with the set are perfectly adequate and do a good job.

A replacement power mike to provide maximum modulation (Courtesy Turner Co., Division of Conrac)

Headsets are available to free both hands and give better modulation (Courtesy Telex Communications)

Basically there are two types of stock mikes—ceramic or dynamic. Either type is fine for the job, although the dynamic type probably rates an edge due to its greater reliability. Most microphones are plugged into the set, either on the front or side. A few sets, mainly less expensive versions, have the microphone wired in directly. The advantage to a plug-in type lies in the ability to replace it with one of the many pre-amp types if you wish. Dynamic mikes are also least susceptible to damage from shock and to extremes of temperature and humidity.

Single Sideband (SSB)

While most CB sets are limited to a maximum of 23 channels, there are some available which have more. This is accomplished by what is called a single sideband (SSB) transceiver.

SSB is relatively new to the CB arena. Basically what it involves is dividing each channel into a carrier and two sidebands, upper and lower, which are duplicates of each other. All single sideband sets are equipped with a mode selector used to select AM, USB (upper sideband), or LSB (lower sideband). In the lower sideband mode, only the lower sideband portion of the carrier is generated, and the carrier and upper sideband are suppressed. In the upper sideband mode, the exact opposite happens, and in the normal AM mode, both USB, LSB, and the carrier are generated. The catch is, that when operating on either USB or LSB, the bandwidth occupied is about 40% of what it would be on normal AM. Therefore, it is possible for two stations to communicate on the LSB of a given channel while two different stations communicate on the USB of the same channel.

Single sideband offers the advantages of greater range, more "channels" (69 as opposed to the 23 on AM) depending on what you call a channel, less interference, and generally better performance. The disadvantage is that the price is higher for SSB equipment and you need another set equipped with SSB to communicate on single sideband. Also, if you go into SSB, make sure that you get a good one or you will find yourself picking up adjacent channels. If you use AM only, at times you might hear what sounds like a quacking duck. It's either that or it's a SSB tranmission picked up on AM.

As previously mentioned, SSB transceivers have a mode switch to switch from AM to LSB or USB. Be careful if you're looking at

Clarifier control (fine tuning) is a "must" for SSB (Courtesy DynaScan Corp.)

an SSB set—some of them are SSB only, meaning that they will not operate on AM. SSB sets also have a fine tuner or clarifier control, which is absolutely necessary for SSB. If it is not set correctly when turned to USB or LSB, the voice you hear may be squawkish, booming, or even unintelligible. It is usually necessary to readjust the clarifier each time a different station is received.

SSB would theoretically permit twice as many CBers to operate on the same channel space, with even less interference than at present. The FCC is presently restudying the entire Citizens Band service, perhaps speculating on the idea of expanding the number of channels, and, at the same time, reducing the amount of interstation interference. But, the basic restructuring of the Class D service, or its expansion would probably take at least 5 years from its inception. So there is really no need to worry that the set you are thinking of buying will be outdated a year later.

MANUFACTURERS' SPECIFICATIONS (AND WHAT DO THEY MEAN?)

Perhaps the greatest confusion in buying a CB rig is the manufacturers' specifications. Nearly everyone knows that there are such things, but hardly anyone knows what they mean. An electronics engineer could tell you real fast what they are all about, but what do they mean to the average buyer? Following are some of the most common specifications and what they mean to you.

Input Power

Input power is usually given in watts, and prior to 1974, transmitter input power was limited to 5 watts. This is actually the amount of power applied to the final amplifier or stage. It does not include the power consumption required to run the transceiver and its various lights and meters. Some manufacturers advertise (and their spec sheets may indicate), that the input power is 5 watts. This is true, but it is the output power which matters to the performance of the set.

Output Power

Output power is also given in watts and applies only to the transmitter. The FCC says that no Class D AM transmitter can put out more than 4 watts of carrier power without modulation. It seems logical, therefore, to look for a set which offers the highest output in watts. But, in actual practice, you will find that most transceivers deliver somewhere around 2½-4 watts, and you will notice little on-the-air difference between 3 and 4 watts. So, don't let this be the only judge.

In most cases, the wattage output is based on the power put into a 50 ohm antenna.

Peak Envelope Power (PEP)

Peak envelope power is a term applied to SSB units only and takes the place of the 4 watt power rating of an AM transceiver. SSB units are also limited to power by the FCC, but in this case the peak envelope power must not exceed 12 watts. The total RF output power of each sideband is 6 watts and the total of both sidebands must not exceed 12 watts.

Modulation

This is where the biggest difference in transmitters is found. FCC rules again limit the amount of modulation to 100%, and the closer to 100% the set is rated, the better its talking power. If there is a choice, get the one which is rated closest to 100% modulation.

Sensitivity

Sensitivity is a rating of the receiver portion of your CB. It is the ability of the receiver to pick up weak signals, and, with the crowded conditions found in most parts of the country, this becomes an important consideration when selecting a set. Sensitivity is normally given on the spec sheet in microvolts (uV), and the smaller the number, the greater its sensitivity. Look for something in the area of under 1.0 microvolt, down to about 0.30 microvolts.

Selectivity

Selectivity is also a rating of the receiver portion and is equally as important as sensitivity. This is the ability of the receiver to reject transmissions on adjacent channels or channels other than that to which the receiver is set. Since signals are separated by small frequency differences, it is important that the selectivity be as good as possible.

The receiver's ability to reject signals on channels other than the one it's tuned to is expressed in dB (decibels) at ± 10 kHz of the channel frequency. The greater the number of dB, the more selective the receiver.

Spurious Rejection

Spurious rejection indicates the ability of your set's transmitter to keep its transmission on one channel. When you talk into the mike, if you over-modulate or put too much talking power into the mike, the undesired portion of your speech can "bleed over" onto adjacent channels.

On the manufacturer's spec sheet, this will be listed in dB (decibels). The better the set, the greater the number, but consider 40 dB an absolute minimum.

Adjacent Channel Rejection

Adjacent channel rejection is similar to selectivity, and is measured in dB (decibels). The larger the number the better, but look for a set with a minimum of 40 dB, but preferably 50 dB or better.

Frequency Range

This is simply the frequencies of the highest and lowest channels at which the set will operate. Class D CB equipment operates at 26.965 MHz-27.255 MHz and this is what should be listed under frequency range.

Frequency Stability

The FCC requires that the set be able to transmit within 0.005%

of the frequency on any given channel. This is the minimum tolerance allowed for the set to be type accepted by the FCC, although it is not uncommon to find tolerances of less than 0.005%.

Squelch Sensitivity

Occasionally, you will find this listed on a specifications sheet. The number given is the most sensitive setting to which the squelch can be adjusted. Putting too much squelch into the audio section will effectively drown out the entire signal, so that there are limits to sensitivity of squelch.

This is typically expressed in microvolts (uV), and the better squelch circuitry gets a lower microvolt count.

Operating Voltage

A base station is designed to be operated from house current, so that the voltage will be listed as 105-125 or 115 VAC. If something such as 60 Hz is given, it can only be operated from a 60 Hz power supply.

Mobile units are designed to be operated from a vehicle electrical system which are typically given as 13.6 VDC, 13.8 VDC, or 12.6 VDC. They all mean the same thing. The only thing you will have to be careful of is polarity. Most mobile sets are designed to be operated from negative ground, and a few offer positive or negative ground operation. Check before you buy.

Current Drain

Current drain is usually a miniscule amount in mobiles and will be given on a spec sheet in milliamperes or amperes. In a mobile you really need not concern yourself with current drain, unless you plan to operate from your vehicle's battery for extended periods.

Current drain on a base set is usually given in watts and is considerably higher than a mobile, due to the power required for tubes and other gadgets not present in a mobile.

Audio Output

You want to look for the highest level of audio output you can find to be able to overcome natural noise. The audio output is the highest level of maximum clear volume which can be attained before distortion occurs; a good audio output rating would be

2.5-3.0 watts. Some of the best (and highest-priced) sets can go as high as 6 watts.

Transceiver Circuitry

This subject could fill volumes and require the education of an electronics engineer to understand. Since all transceivers have to meet certain standards to be type accepted by the FCC, they are all fairly similar to those unversed in electronic circuitry. Unfortunately, there are also many variations.

There are, however, two basic types of circuitry in use today; single conversion receiver and double conversion receivers. Both are of the type known as superheterodyne. Without getting into a long electronic discussion of the differences, it all comes down to this: a double conversion superheterodyne improves its selectivity and gives itself a better chance to reject off-frequency signals. But don't by-pass a set just because it is not a double conversion receiver. The selectivity of a single conversion receiver can equal that of a double conversion if the selectivity filtering is good enough.

Most transceivers use crystals such as these to control frequencies

Almost all receivers used to be controlled by crystals, as were all transmitters. Each channel required two crystals; one to transmit and one to receive. This adds up to 46 crystals for a 23 channel set. Efforts at cost reduction have produced what is known as "crystal synthesis," a means by which considerably fewer crystals are used for full 23 channel coverage, saving approximately $50.00 per transceiver.

While almost all 23 channel sets today use crystal synthesis, there are a few models using synthesis in user-selected channel models. The advantage here is that only one crystal need be added or changed to change a channel capability.

Knowing what to look for, you'll find the selection enormous. Try to choose right the first time, but if you don't, there's a good market for used sets. CB is a "move-up" market. It's like boats—you must have a bigger, better one, or in this case, one that gets out further. After you buy the first one, you'll known exactly what you should've bought in the first place.

TYPICAL MANUFACTURERS' SPECIFICATIONS

SPECIFICATIONS GENERAL:

Channels:	23 Crystal-Controlled
Size:	6-3/4″ Wide 2-1/4″ High 8″ Deep
Weight:	4.2 Pounds
Antenna:	52-Ohm Coaxial
Primary Power:	Input Voltage—13.8 VDC

RECEIVER:

Frequency Range:	26.965 MHz.—27.255 MHz.
Sensitivity:	0.3 μV for S +N/N using 1,000 Hz. 30% Modulation
Selectivity:	6 db bandwidth 5 KHz. 50 db bandwidth 20 KHz.
Cross Modulation:	75db for 10 μV desired
Spurious Rejection:	60 db minimum
Adjacent Channel Rejection:	50 db minimum
Squelch Range:	Adjustable from 0.5 μV—1,000 μV
Noise Limiter:	ON/OFF
1st I.F. Frequency:	10.635 MHz. for center frequency
2nd I.F. Frequency:	455 KHz.
P.A. Maximum Audio Output Power:	5 W
Audio Output Power for 10%:	3.5 W
Speaker:	3-5/8″

TRANSMITTER:

Frequency Range:	26.965 MHz.—27.255 MHz.
Carrier Frequency Stability:	0.003%,−30°C to +65°C
Output Power:	3.2 W into 52 ohm with 13.8 V DC power supply
Modulation Capability:	100%
Spurious & Harmonics Suppression:	55 db minimum

THE ANTENNA SYSTEM

The antenna that you will use with your CB is going to determine your rig's performance, for this is the most important part of the rig. No matter how expensive or how good your transceiver is, it will not give peak performance unless it has a good antenna capable of radiating and receiving a strong signal. A poor antenna or poor antenna installation can make a $300.00 transceiver work like a piece of junk, just as a less expensive transceiver will work beyond expectations with a good antenna.

Because of the low power limitations placed on CB equipment, a good antenna becomes even more important. Normally, even though you have about 5 watts of input power to the transmitter, the set is only supplying about 3½ watts to the antenna and the output may even be less, depending upon the circuitry, transmitter efficiency, and energy required to run various lights and meters. In the case of small hand-held "walkie-talkies" the power may even be less, down around a fraction of 1 watt. However, optimum performance can be obtained from almost any set by carefully selecting and installing an antenna system which will be capable of using the limited power available to the best advantage. This includes not only the antenna, or aerial, but also the coaxial cable used to connect the antenna to the transceiver, various connections, and any matching devices which may be used. Failure of any part of the system will affect the performance of the entire system.

Many times a perfectly good transceiver is blamed for an inability to transmit effectively, when in fact, it may be the fault of the antenna system, which is the case in most instances. It is fairly easy these days for manufacturers of CBs to reach the legal design limits imposed by the FCC. The communicating range, therefore, lies in matching the antenna to the transceiver, for if the transceiver output remains a constant, it is possible to raise the communicating range of the rig simply by improving the antenna efficiency. An omnidirectional antenna used by almost all mobiles, can only radiate the amount of energy supplied it by the transmitter, but newer, more sophisticated base station antennas have a signal gain. This is accomplished through the ability to point or "aim" radio signals, making them appear to be much stronger than they actually are. A "gain" type of antenna can make 3 watts seem like considerably more as it leaves the antenna and can result in better overall performance simply by utilizing improved antenna efficiency.

TYPES OF ANTENNAS

It is not always possible to select an antenna that you want. Small walkie-talkies usually come equipped with a telescoping whip antenna, although more expensive walkie-talkies have provision for using a small portable-type antenna which fastens directly to the set with a standard PL-259 connector. These are also capable of using remote outdoor antennas, but this is rare. Twenty-three channel mobile and base sets almost always use remote antennas.

CB antennas come in all shapes, sizes, and prices, but there are basically whips, ground planes, coaxials, and beams, each with its myriad variations.

Walkie/talkie telescoping antennas can usually be replaced with clip-on loaded whips or with one-piece rubber whips.

WHIPS

More commonly known as a vertical whip, it is used mostly for mobile installations. It derives its name from its whipping action, or flexibility, and is normally ¼ wavelength. A whip can be either base, center, top or continuous-loaded, depending on construction, to reduce its physical length and is available in steel or fiberglass. Marine antennas are generally of this type because of the abuse taken from the elements and from normal wear and tear on equipment.

Some of the advantages and/or disadvantages to whips are:

CONTINUOUS LOAD:	Easier to install, but less critically tuned
TOP LOADED:	Most efficient, but harder to match. Their radiation characteristics frequently change due to the tendency of top weighted antennas to sway. The most fragile part (coil) is at the most vulnerable spot (top).
CENTER LOADED:	Efficient radiation patterns, but the weight causes a lot of sway.
BASE LOADED:	Offer slender profile and less susceptible to picking up interference generated by passing vehicles. They are also more tolerant of the low capacitances between car and ground.

COAXIAL

Base stations are the primary users of coaxial antennas, although they find occasional use with mobile rigs. These provide a radiation pattern which is omnidirectional, radiating evenly in all directions. They are popular, requiring little space, are available in ¼ wave (mobile), or ½ wave (base), and can be modified to provide a certain amount of gain, omnidirectionally.

GROUND PLANE

Ground planes are by far the most popular in CB use. In its basic form it is omnidirectional and provides no gain, but in any one of its myriad variations, can give an omnidirectional gain or even be "aimed" to a slight degree. While ground plane antennas were designed primarily for base station use, they have found a large application in mobile use, using the roof or trunk lid of a vehicle for a ground plane.

BEAM

Beam antennas are extremely directional (radiating in one general direction only), but capable of providing high power gains. By arranging the elements properly and spacing them at proper intervals, it makes the radiating signal add up along a single direction, forming a "beam" (hence the name). Power is thereby gained, but in only one direction, and the more directional, the higher the gain.

This arrangement is ideal for fixed station point-to-point communication, but for everyday application, the beam antenna must incorporate a rotor to turn the antenna and aim it in the desired direction.

ANTENNA FEATURES

Almost all two-way radio equipment uses the same antenna for transmitting and receiving. Switching within the transceiver is used to connect the antenna to the transmitter output side or the receiver input side. But not just any old hunk of wire will do for an antenna. The AM/FM car radio antenna is simply not suitable for a CB antenna, although there are several good combination AM/FM/CB antennas on the market. These types are tuned for optimum CB performance and will work just as well with the AM/FM receiver. About the only CB recievers which use the car radio antenna are several models of converters which convert the AM/FM receiver into a CB receiver.

Antenna Length

There is a direct, scientific relationship between the physical length of an antenna and its electrical properties. An antenna will only function properly and efficiently when its own physical length is in a mathematical relationship with the wavelength of the signal which it is supposed to receive or transmit.

The wavelength of a radio wave, or the distance which it will travel in one complete cycle, can be calculated by dividing the velocity of the radio wave by its frequency, or number of hertz (Hz). Hertz means cycles per second and CB operates at approximately 27 MHz (27,000 hertz or cycles per second), and a CB radio wave will travel about 11 meters in one cycle. One meter is 39.37 inches, so 11 meters are:

$$39.37 \times 11 = 433.07 \text{ inches (36 ft. 1 in.)}$$

Since an antenna will not operate efficiently unless its physical length is equal to, or a multiple of, the wavelength at any given frequency, a CB antenna should theoretically be slightly over 36 feet long.

This is obviously the most efficient antenna, and the most impractical, so that most CB base station antennas are cut in half, to about 18 feet, and called half-wave antennas. They are essentially the same as a full-wave antenna, but far more practical.

But this still leaves the matter of mobile antennas. Even a half-wave, 18 foot long antenna hanging off the bumper is a precarious situation. So, for mobile applications, the ¼-wave antenna is far more suitable, presenting less of a danger to bridges, gas station lights, and low-flying birds. The ¼-wave antenna is only about 108 in. long (433.07 ÷ 4 = 108), also known as a "108 whip."

Loading Coils

For mobile CB application, people do not always want even 108 inches of fiberglass or stainless steel hanging from their car or truck bumper. The solution to this problem is to use an antenna which incorporates a loading coil.

The theory behind a loading coil is that, while using a physically shorter antenna, you can "fool" the antenna into thinking it is the proper length, usually ¼-wave or ⅝-wave. These antennas come anywhere from slightly under 18 inches long to around 4-6 feet long, with loading coils which are actually wire windings to increase the eletrical length of the antenna.

Coils are either base, center, top, or continuous-type, and are used primarily on mobiles. The position of the coil on the antenna determines its type, with a continous load done by spiral winding the conductor along the entire length of the antenna.

These are the most popular with mobile rigs because of their small size and the fact that they usually can be removed when you are parked in high-theft areas, such as parking lots, and are also less an obstacle in campgrounds, gas stations, or out on the trail. They do, however, present the added complication of sometimes having to drill a hole somewhere on your vehicle to install them, rather than mounting them on the bumper, although mounts are available requiring no holes.

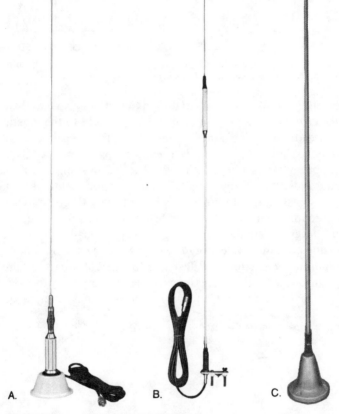

A. Base loaded antenna (Courtesy Breaker Corp.)

B. Center loaded whip (Courtesy Breaker Corp.)

C. Continuous loaded whip with marine deck mount (Courtesy Midland International)

Antenna Gain

Antenna design will determine radiation pattern which, in effect, will determine "gain." Antenna gain is the alteration of the radiation pattern, concentrating the radiated energy in one direction to make it appear that it was produced by a much stronger source. This does not mean that the radiated signal was actually amplified.

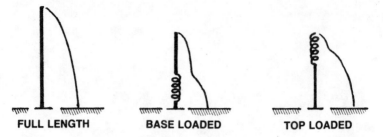

FULL LENGTH   **BASE LOADED**   **TOP LOADED**

An antenna can be "fooled" into thinking it is electrically longer than it actually is with a loading coil. (Courtesy Shakespeare Industrial Antenna Div.)

In fact, all that took place was to design the antenna so that it was able to concentrate a portion of its energy to add to the existing signal already following a desired direction. The db gain is determined by just how much energy is directed along this desired path.

A mobile whip is normally referred to as an omnidirectional antenna, radiating energy equally in all directions, but achieving minimal signal gain unless specifically designed to do so. These omnidirectional gain antennas are designed to lower the angle at which energy is radiated, diverting energy normally lost as sky waves into ground wave energy producing a gain in all directions.

A unidirectional antenna is designed to concentrate its radiated energy in one direction. This means that by applying the same power to a unidirectional and an omnidirectional antenna, the

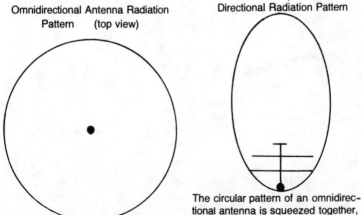

Omnidirectional Antenna Radiation Pattern   (top view)

Directional Radiation Pattern

The antenna will radiate the signal in all directions

The circular pattern of an omnidirectional antenna is squeezed together, forming an oval shaped pattern, giving greater distance in one direction and sacrificing it in other directions.

unidirectional antenna will deliver more energy in a specific direction, but at the expense of field strength in other directions. A unidirectional antenna radiation pattern will cover the same area as an omnidirectional, but will take power from other directions and direct it to another area. This is why unidirectional antennas are mostly used for base sets where they can be equipped with a rotor to point them in the desired direction.

Use an antenna with no more than 6 dB gain, especially for marine use. Any more then 6 dB gain will squeeze the signal together and affect the radiation pattern. Because the radiation pattern is too concentrated, as a boat pitches and rolls, the signal can be powered right over a receiving antenna.

GROUND WAVES AND SKY WAVES

No discussion of antennas and their characteristics would be complete without mentioning ground and sky waves. Radio energy leaving an antenna consists of both ground and sky waves, which, as they travel through the atmosphere, are bent or broken up. The frequency, antenna type, and atmospheric conditions are only a few of the factors which can influence the travel of radio waves.

Ground waves generally follow the curvature of the earth and will usually allow communications with stations located over the horizon, depending largely on terrain and presence of large buildings, etc.

Sky waves follow an upward path into space from the antenna. Depending on the angle at which they encounter the ionosphere, they can either be bent or reflected back to earth as far away as a thousand miles. This particular phenomenon is known as "skip." It can occur once or more than once, depending on conditions, accounting for transmissions received at great distances.

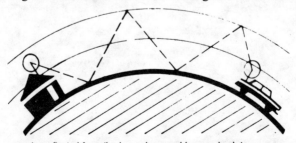

Skywaves can be reflected from the ionosphere and bounce back to earth thousands of miles away. (Courtesy Linear Systems, Inc.)

The "skip" characteristics of the 27 MHz citizens band are most effected by sunspots activity, which varies during an 11 year cycle (and is currently at an interference low) and the time of day. "Skip" signals are capable of traveling thousands of miles, but may fade out in a few minutes, and are essentially a daytime phenomenon.

RADIO AND ANTENNA INSTALLATION

Almost all CB equipment that you will purchase new comes with some sort of owner's manual, set-up instructions, or installation procedures. Installation is not very hard, although in some cases (base sets), it may require some crawling around on the roof to get the antenna as high as possible. Follow the manufacturer's instructions and supplement them with some ideas here and you can do just as good a job as some "self-styled experts." The hardest part is buying the right pieces to do the job.

Some common hand tools are necessary: electric drill and bits, screwdriver, pliers, soldering gun and resin core solder, assorted wrenches, and wire strippers, etc. You will also need some wire end terminals and a few female spade-type connectors if you are going to go to the fuse block of your vehicle to power your mobile rig.

INSTALLING MOBILE CBs

Mobile installations are not hard. The foremost problem is deciding where to put the set so that it will fit neatly, be relatively unobtrusive, be within easy reach, and be the least hassle to install. Each factor should be weighed before deciding on a place.

Most sets these days do not present much of a problem because of their small size. Many could fit inside the glove compartment, except that they are hard for the driver to get at. The transceiver should be placed within easy reach of the driver so that operating the set is not distracting. The central engine cover on vans is a natural place, as is the below dash mount in passenger cars. Kits are also available for vans to provide a headliner installation, and one company offers an overhead console. These type installations have the advantage of the speaker facing downward and giving better sound. Other alternatives include transmission hump mount

47

In dash mount (Courtesy J.I.L. Corp. of America, Inc.)

brackets, in-dash installations, or practically anywhere your imagination deems best.

Motorcycles usually mount the CB on the gas tank in a cradle supported on rubber dampers, as this is about the only possible place. RVs and tractor-trailers usually mount the radio on the lid of

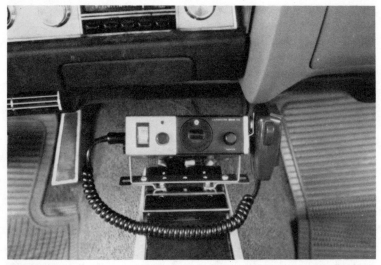

Console mount

Above dash mount

Below dash mount

the dash console or on the windshield crossbeam preferred by 18-wheelers. Boaters normally mount the unit on the dashboard or on the dash gauge console.

The unit should fit neatly and unobtrusively, with emphasis on the unobtrusive. This is not to say that a careless installation is desirable, but CBs have become a popular item with thieves. Unless you plan to take your set with you or disconnect it every time you leave your vehicle, you had better give some thought to installing it where it will attract the least attention. Several other

Engine cover mount for a van or motorhome (Courtesy Kris, Inc.)

Some companies offer overhead mounting consoles (Courtesy J.I.L. Corp. of America, Inc.)

options are available here. There are CB lock mounts available to padlock your set to the car, but thieves are generally not too subtle about their work. Most times they will take the CB, lock mount and all, or worse take the whole car. A fairly good solution is to use the slide mounts popular with tape deck enthusiasts to mount your radio. This way you only have to disconnect the antenna lead-in and slide the unit off its mount to stow it in the glove compartment or under the seat. Another simple, but slower, solution to the problem is to use a two-pronged trailer harness to provide a quick-disconnect means for the power and ground leads to your set. After uncoupling the quick-disconnect, simply disconnect the antenna and the 2 or 4 screws holding the set to the bracket and stow the set out of sight and mind.

CB designed specifically for a motorcycle, held in place on the gas tank by straps. (Courtesy Beltek Corp.)

A slide mount will facilitate easy removal of the set

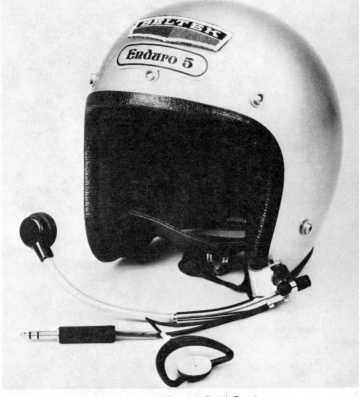

Headsets are available for bike helmets (Courtesy Beltek Corp.)

After spending many agonizing hours deciding where to put your new set, here's how to do it.

• Open the box and lay out all the pieces. Study the instructions and make sure that you got everything you paid for, and that you have a mental idea of just how the thing will be installed.

• After satisfying yourself that you have found the ideal mounting place, and you are sure it will fit there, attach the bracket to the radio and hold it up in position. Check that the antenna cable can be easily connected or disconnected, and that the radio does not interfere with any heater or dash-mounted controls. Mark the position of the bracket on the sides or the forward edge.

If you have decided to use a slide mount for easy removal, you may or may not have to mount the normal bracket on the slide mount. It depends on what kind of set you have and what type of

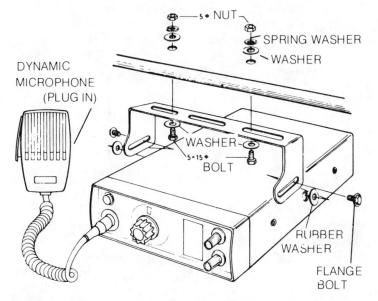

Typical under dash installation. To mount the set on top of something, install the mounting bracket upside down. (Courtesy Sparkomatic Corp.)

slide mount you have. In any case, follow the manufacturer's instructions for installing the slide mount. It is not really different from installing a normal mounting bracket, except that one piece attaches to the vehicle and one to the radio. Electrical contacts are provided so that when the slide mount is engaged, electrical contact is made.

• Take the bracket and set down, and remove the bracket. You're ready to mark and drill the bracket mounting holes. Put the bracket back in place and align it using the marks you just made. Mark the position of the center of the holes with a pencil dot. If you're afraid of scratching your custom paint job with the drill, lay down some small pieces of masking tape prior to marking the holes. Using a center punch and hammer, centerpunch a small hole where the pencil mark is. This will prevent the drill from scratching surrounding metal as you drill the hole. Use a drill stop on your drill bit to avoid drilling through the mounting area when you don't know what's behind it. For most installations, use a drill which is the same diameter as the root diameter of the screw.

• Using the screws or the bolts and nuts provided, securely fasten the mounting bracket in place. If you don't like the screws

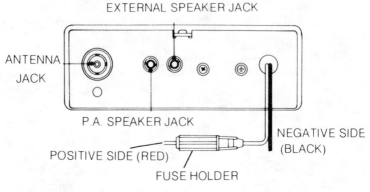

EXTERNAL SPEAKER JACK

ANTENNA JACK

P.A. SPEAKER JACK

POSITIVE SIDE (RED)

FUSE HOLDER

NEGATIVE SIDE (BLACK)

Typical connections at the rear of the set

they gave you, get some you do like. Boats with fiberglass dashboards, and similar installation, will requre some longer bolts with nuts and lockwashers and backing of the mounting area with a piece of plywood to distribute the weight of the unit. Install the radio in the bracket using the slotted screws or thumbscrews provided. Some sets also use cushioning rubber washers between the set and the bracket. If you plan to remove your set from the bracket frequently, glue the rubber washers in place or they will likely become the subject of abusive language. It is no fun trying to hold a rubber washer in place and fit a screw through it in the dark.

• Connect the microphone to the set, and hold the microphone at various places until you have found the best mounting spot. The mike hanger serves two purposes: (1) it keeps the installation neat and the mike out of the way when not in use; and, (2) if the microphone is allowed to hang from the set, it will stretch the cord and connections. Generally, you'll want to mount the mike in a convenient position where, eventually, you won't have to look to

Magnetic mike holder

hang it up; you'll do it by feel. Likewise, you don't want it in a space where the cord is in the driver's way. You can screw the mike hanger to the dashboard using the same procedure for installing the bracket, or remove one of the screws from the radio housing and attach the hanger to the side of the set. If the idea of drilling more holes in your dash is repugnant, you can use a magnetic mike hanger.

• The next step is to determine battery polarity. You should have done this before you bought the set, but if you didn't, you're probably lucky (or prudent) enough to have bought a negative ground CB set. There are a few sets available which will work with negative or positive ground depending on how they're hooked up, but the vast majority are for 12 volt negative ground systems. If you are working with a 6 volt system, power inverters are available to convert 6 volts to 12 volts.

Most cars, trucks, RVs, motorcycles, and boats use a 12 volt negative ground electrical system, although some heavy trucks are positive ground, and some older vehicles are 6 volt, positive ground. The battery polarity can be easily identified by seeing which battery terminal is connected to the chassis for ground. This is the ground side and the other side is the "hot" side or power source.

CAUTION: *Never, ever, hook a negative ground set to a positive ground or vice versa. Be sure to check polarity before hooking up a CB.*

• For most negative ground installations there are three popular sources of power, to which the fused lead on the CB set can be connected.

The best source is the positive, or the "hot," side of the battery. This is the best and most stable source of voltage, will lead to the least noise interference, and will provide power whether or not the ignition is "ON." Installations on boats and motorcycles will frequently require this power source. A second source of power, and probably most popular, is the fuse block, usually located under the dashboard. Tapping the fuse block for power will provide a relatively stable voltage source and will enable your set to operate independently of the ignition switch, depending on which accessory terminal of the fuse block you tap. If this method is used, you will need to solder a female spade connector onto the end of the transceiver "hot" lead. A third, and least popular, way of getting

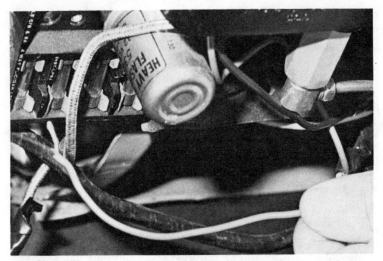

The automotive fuse block provides a popular and convenient power source. (Courtesy Pace Communications, Div. of Pathcom, Inc.)

power to the set is to use the accessory lead or terminal on the ignition switch, which will enable operation of the set only when the ignition switch is in the "ON" or "ACC" position. This method is rarely used since the advent of the steering column-mounted ignition switch.

To tap the cigarette lighter for power, attach the ends of the adaptor to the transceiver leads. Be sure to observe proper polarity. (Courtesy Pace Communications, Div. of Pathcom, Inc.)

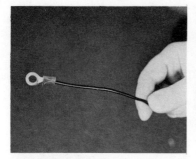 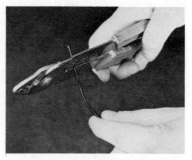

Electrical ring lug installed   Wire strippers avoid cutting into the wire

Whichever method you select, be sure that you have the right polarity and be sure that the fuse clip (usually supplied with the set), is installed between the transceiver and the power source in the "hot" lead. Consult the manufacturer's recommendations, but the transceiver "hot" lead is generally fused with a 1.5 amp fuse, available at any automotive store.

Cigarette lighters adaptor power cords work fine, too, but simply tapping other "hot" wires is not a recommended power source for best performance, as they rarely provide a stable voltage source and can be the source of interference.

Once you've decided on a power source, you will have to install a suitable terminal on the end of the "hot" wire. For marine use, you can simply attach an electrical ring lug big enough to go over the battery terminal. For fuse box operation, install a female spade lug.

• It is important when installing terminal hardware on wire ends that you get a solid connection, or noise can enter the radio at this point. Cut the wire to length and leave a little slack. It's easiest to use a pair of wire strippers to strip about a ½ in. of insulation off the end. Use the groove in the strippers corresponding to the gauge wire you're using. Twist the loose wire strands together tightly and insert them into the lug so that the insulation butts against the barrel of the lug. Crimp the lug around the wire securely, using either the crimpers on the end of the pair of wire strippers or an ordinary pair of pliers. If you use pliers, try to get one side of the lug barrel to go under the other, making them overlap. When this is tightly crimped, finish the job by soldering the connection, after trimming the wire end which protrudes from the lug barrel.

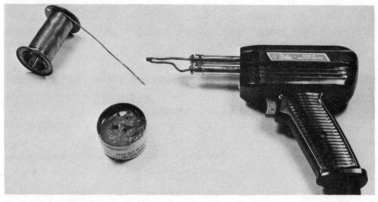

Be sure you solder all connections to make them tight and trouble-free

If you're using a slide mount, the same procedures apply for wiring the stationary part of the slide mount as for wiring a CB. The only difference is that you'll have to follow the manufacturer's instructions for wiring the CB to the movable part of the slide mount. Strip a little bit of insulation away from the wire and solder it to the contacts. Make sure that both parts of the mount are using the upper and lower sides of the same contacts, or you'll have no electrical contact.

• The "hot" lead is almost ready to be connected. Before going any further, cut the "hot" lead and install the fuse holder with fuse. After twisting the wire together in a "pigtail" connection, drop some solder on the pigtail and wrap it neatly with electrical tape.

• Decide where you are going to ground the transceiver. This can be almost anywhere—a screw or bolt which is nearby and electrically connected to the frame. If you can't find a screw

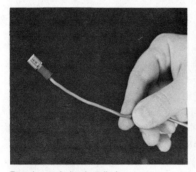

Female spade lug installed

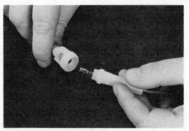

All transceivers should have an in-line fuse installed between the set and the power source. (Courtesy Pace Communications, Div. of Pathcom, Inc.)

The ground lead from the transceiver can be connected to a convenient bolt or screw. Use a toothed washer for good contact.

nearby, drill a hole and use a sheet metal screw, in an out-of-the-way place (for neatness). Crimp and solder a ring or open-type lug onto the ground lead after cutting it to length. Use the same procedure as with the "hot" lead.

If your CB is not equipped with a plug on the rear of the set to disconnect the power and ground leads, you can do this easily. Most automotive stores sell 2-prong trailer connectors which only go together one way. They are inexpensive and solve the problem of disconnecting a ground lead from a bolt every time you want to remove your set from its mount. The 2-prong connector can be spliced into the power and ground leads easily, making sure that

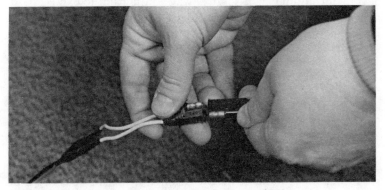

A power lead "quick-disconnect" can be homemade for under $2.00

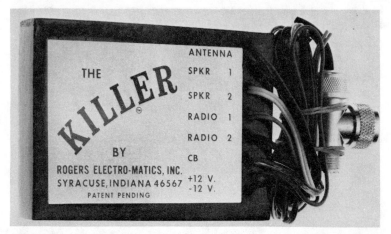

Accessories such as this one are available to allow simultaneous CB and AM or FM radio operation. When a signal is received on the CB set, strong enough to break the squelch wherever it is set, it will cut out the speakers on the AM or FM radio. When no CB signal is received, the AM or FM radio will automatically return to normal operation. (Courtesy Rogers Electro-Matics, Inc.)

the circuits maintain their continuity. The connectors are generally color-coded to make this simple. Pigtail the wires together, apply a drop of solder, and securely tape the connection for a quick and easy disconnect.

• Turn the set "OFF" and route the leads. If it is necessary to route either of the leads through any sheet metal, drill an oversize hole and use a rubber grommet to prevent chafing. Stuff the leads out of the way and secure them with plastic ties salvaged from the kitchen. Connect the "hot" lead and the ground lead. When connecting the ground lead, scrape away a little paint to expose bare metal and use a toothed washer to get a solid ground. Tighten the grounding screw securely.

WARNING: DO NOT, UNDER ANY CIRCUMSTANCES, OPERATE THE TRANSCEIVER WITHOUT CONNECTING THE ANTENNA. YOU COULD BLOW OUT THE OUTPUT TRANSISTOR.

The installation of the set is complete; all that remains is to install the antenna, and give some thought to thievery.

INSTALLING MOBILE ANTENNAS

It would be impossible here to detail how to install each type of antenna with each type of mount which is available; there are probably more antenna and mount combinations than there are brands of CB sets. Before actually going out to get an antenna, look over your car, boat, or truck and decide the best place for your antenna, for this will go a long way to determining what kind of antenna you get. Try a large CB suppliers catalog—there is a mount for every purpose. If you have a car or pick-up truck, you can use just about any kind of antenna available—mirror mount co-phase, cowl mount, bumper mount, rooftop, or trunk lid mount. A base-loaded trunk lid mount is probably most popular with passenger cars because it requires no hole drilling in the body. In popularity, these are followed closely by 102 inch whips on bumper mounts and center loaded antennas on rain gutter mounts. More often than not, a van will have a base-loaded antenna installed on the roof to take advantage of the large metal surface area, affording a good ground plane. Pick-up trucks and tractor-trailer operators seem to favor the single or dual mirror mount, usually center-loaded. The reason is that the large expanse of metal on trucks will do a good job of blocking the radiated signal pattern and reflecting a good bit of it back into the antenna. The center-loaded antennas get the radiation pattern up in the air, away from sheet metal. Boats usually are equipped with specially-designed marine antennas,

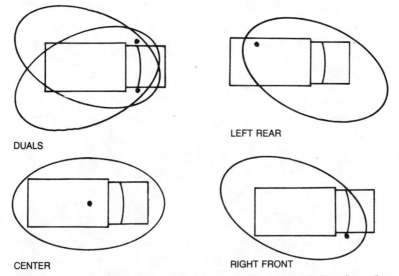

DUALS

LEFT REAR

CENTER

RIGHT FRONT

Effects of antenna location on range and direction. This will also vary slightly depending on the vehicle

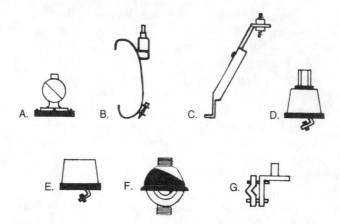

Most of the popular antenna mounts—(A) swivel ball, (B) bumper mount with ball, (C) trunk groove, (D) trunk lip, (E) trunk groove for snap-in antenna, (G) mirror mount and (H) cowl mount adaptor for base loaded antennas (Courtesy Hy-Gain Electronics Corp.)

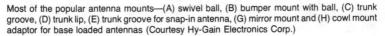

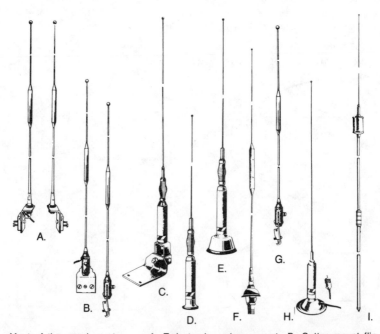

Most of the popular antennas: A. Twin trucker mirror mount B. Gutter mount flippers C. Base loaded for a motor home D. Base loaded snap-in roof mount E. Base loaded trunk mount F. Cowl mount center load G. Center load gutter mount flippers H. Temporary magnetic mount base load I. Bumper mount center load although they can use almost any kind of base, center, or continuous load.

The basic consideration in antenna installations is height, to get the antenna as far off the road as practical or possible. Other considerations include location (it will affect the directionality of the set), and matching, which can be accomplished easily.

The possibilities for mounts and antennas are endless. They run from the simple gutter mount, to the popular trunk lid mount to the co-phased dual mount, known as "twin truckers." Some require hole drilling, some do not. Some can be equipped with quick-disconnect arrangements, others cannot. Naturally, two antennas are better than one, but be warned: they must be at least 78 inches apart or no better performance will result. Some mounts will give better performance than others. You have to be very careful when mounting antennas on pick-up truck bodies or on West Coast-style mirror brackets. Either of these items can easily become electrically isolated from the chassis ground, resulting in an open circuit.

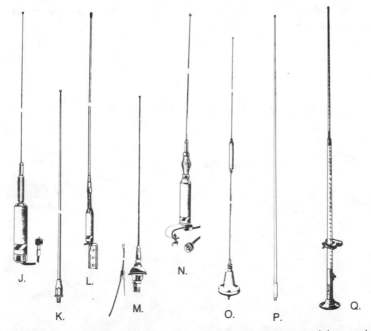

J. Center load K. Continuous load L. Non-ground plane (requires no ground plane, can be mounted on any surface) M. AM/FM/CB cowl mount N. Quick grip base load O. Marine deck mount center load P. 102 inch whip Q. Heavy duty marine deck mount

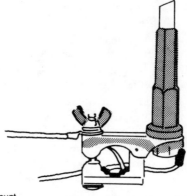

Temporary mirror mount

It's not an easy task to select an antenna; you'll probably agonize over it for hours, trying to balance off all the factors: maximum performance, location, type of antenna, cost, type of mount, length of cable, etc., etc. If at all possible, buy an antenna system, which is, one that comes packaged with the type of mount and antenna you want, the proper length of cable for the antenna, and the hardware. This will save having to buy the pieces separately. Also look for an antenna which has a resonator tip, tunable either by cutting it or by moving it up or down. This will save considerable time and effort.

• Once you have purchased the antenna system (or the antenna and the necessary hardware), lay out all the pieces and see how it

Cowl mount installed

Trunk mount installed

goes together. You'll have to rely on the instructions packaged with the system, since there are too many variations to detail each one here. If you bought your antenna system piecemeal, you're basically on your own, except for the recommendations made here which are common to all antennas.

• If you have a co-phased (dual) system, the place to start is at the transceiver by screwing the PL-259 connector into the set. All other installations should begin at the mount.

Do not shorten the coaxial cable supplied with dual antennas. Co-phase cable is RG-59/21 of 72 ohms impedance (net 50 ohm) and single whips use RG-58/21 of 50 ohm inpedance. In all other installations, a general rule of thumb is to use the shortest cable possible, although it is not a wise idea to cut the length of the cable supplied with your antenna. It is usually around 18 feet long to accommodate trunk lid mounts, and should be used as it is supplied.

• Install the antenna mount on the vehicle. Remember, marine antennas should be placed on the opposite side from the engine control panel. If you are using a base-loaded antenna with a "no-hole" mount, it is relatively simple, as are bumper mounts. A ball mount or cowl mount will take a little more time. If it is necessary to drill a hole for the mount, drill a small pilot hole first and gradually enlarge the hole with several bits. From here the hole can be reamed to size (after applying a circle of masking tape around the hole to prevent paint chipping), or a metal hole saw can be used. Either way, file the edges of the hole when you're done to remove jagged edges.

• Connect the antenna cable to the antenna at the mount. Depending on the type of mount you're using, there are several ways

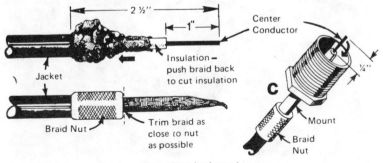

One method of connecting the cable to a base load mount
(Courtesy The Antenna Specialists Co.)

of doing this. Most base-loaded antennas use a screw-on type mount. The outside insulation is stripped off to about 1 inch back, being careful to leave the coaxial shielding intact. Push the coaxial shield back and carefully remove the foam insulation from the center conductor. Slip the bare center conductor up through the hole in the mount and bend it over in the channel provided. Ground the coaxial shield as specified.

Ball mounts, mirror mounts, bumper mounts and marine deck mounts, usually require that lugs be soldered onto the ends of the cable. Carefully cut away the outer insulation for about 2 inches, leaving the coaxial shield intact. Unbraid the coaxial shield and twist it together. Use the same procedure for attaching a lug of sufficient size, as was used to solder a lug onto the transceiver leads (also in this chapter). Cut away about ½ inch of the foam insulation from the center conductor and attach another lag to it in the same manner as before.

Connect the antenna and ground leads as suggested by the manufacturer. Clamp the wires in place to avoid chafing. Be sure to use all the insulation pieces supplied, and in the proper order of assembly.

• You are ready to route the cable. On trunk lid mounts the cable can be routed through the trunk (leaving enough slack to open the trunk), under the rear seat and along the driveshaft tunnel, coming up under the transceiver. It can also be routed along either side

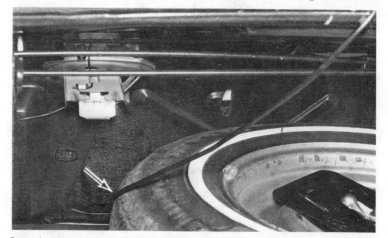

On trunk mounts, route the cable through the trunk leaving enough slack to open the trunk. (Courtesy Pace Communications, Div. of Pathcom, Inc.)

From the trunk, or after bringing the cable through the roof or in the window it can easily be routed under the door sill threshold plate (Courtesy Pace Communications, Div. of Pathcom, Inc.)

under the door sill threshold plate. Co-phased harnesses should be routed under the dash and through the body, using a grommet. Sometimes it is easier to route it through each door, since the mirrors are usually on the door. Be sure to leave enough slack in the cable to allow doors to open. For mirror mounts, run the cable along the mirror brackets and secure them with toothed plastic bundling ties. Connect them to the antenna mounts as previously described.

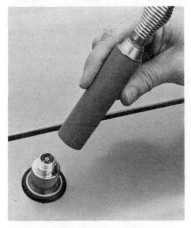

Install the antenna on the mount. Do not over-tighten base loads

Cowl mount antenna installed

Cowl mounts usually require that the cable run through the firewall at some point. Try to route it away from the fuse box, electrical accessories, etc. Drill an oversize hole and use a grommet to prevent chafing.

Excess cable should be routed so that it is not in a tight loop. It should be in free, loose coils to prevent picking up interference.

• Install the antenna on the mount. At this point you can use any springs you might want, which should be installed now. Fiberglass antennas usually require a spring of some sort to prevent cracking when they hit low tree limbs, etc. Slip a piece of surgical tubing or vacuum hose over a long whip, where it could possibly chafe the body.

A quick disconnect is a good idea at this point, too. It will allow you to push down and turn the antenna to pick it off the mount any time you desire. Base-loaded antennas don't need this as they simply unscrew from the mount.

TUNING THE ANTENNA

Now that everything is installed and the vehicle is back together, it only remains to tune the antenna. It must be tuned to obtain the lowest possible VSWR (variable standing wave ratio), which is a measure of how much radiated signal power is reflected back into

The resonator tip on a base load is held in place by a setscrew

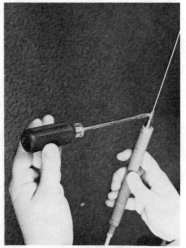

Center loads also use a setscrew to hold a much shorter resonator in place

the antenna. Most antennas are factory tuned for 1.5:1 or better, but VSWR of 1.1:1 is considered ideal, but seldom attained, and VSWR of 3:1 or more damage the transmitter.

Antennas are tuned in several ways:

A. Base-Loaded: These are usually equipped with an allen head set screw which is loosened to move the antenna shaft up or down. It may even require a little filing to shorten it.

B. Center-Loaded: These types usually have a resonator tip made of wire held in place by a set screw. They are also tuned by moving the resonator tip up or down or even shortening it slightly.

C. Continuous-Loaded: Continuous-load antennas are tuned either by loosening a thumbscrew and sliding a resonator tip up or down, or by removing the little cap from the antenna and filing a slight amount off the top.

D. Some antennas are tuned by sliding a small ball up or down the resonator.

Check the instructions before attempting to tune your antenna. You will need to buy (about $15-20), or borrow a SWR meter, and maybe a 2 foot long patch cord equipped with a PL-259 on each end, or install a CB match box, to match antenna and transmitter.

Antennas should be tuned away from tall buildings, preferably in a wide open space, such as a parking lot on a day that the store's

You'll need an SWR meter to trim the antenna

CB matchbox to match antenna and transmitter (Courtesy Midland International)

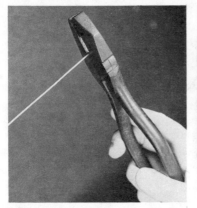

Resonator tips on center loads can be cut with a pair of sidecutters, but file the end square when you're finished. Base load resonators usually have to be ground off as they are much thicker at the base.

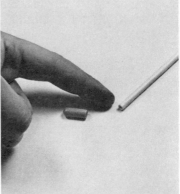

Most fiberglass antennas can have a slight bit of the tip filed off to tune them. Remove the plastic cap at the tip

closed. Turn off the set and connect the SWR meter between the antenna and the transceiver; you will see an end marked "ANTENNA" and one marked "TRANSMITTER." The front of the meter has a switch, one side marked "FWD" (forward) and the other marked "REF" (reflected). Be sure that all doors and trunk lids are closed. Set the transceiver to Channel 1 and flip the switch to "FWD" and key the mike. The needle will jump. The SWR meter also has a "zero" knob. As you key the mike with the switch in FWD position, turn the "zero" knob until the needle aligns with the calibration line on the meter face. Let go of the mike switch and flip the meter to "REF." Key the mike again and read the SWR. Anything 1:1.5 or less is good and you need not go any farther. The other channels can be checked in the same manner. Just be sure you "zero" the meter each time you check.

The lower-numbered channels will generally read a slightly lower SWR than the higher-numbered channels. An ideal match would have Channel 12 at 1:1 and channels 1 and 23 at 1.2:1. The antenna is better slightly longer than too short.

If the SWR is above 1.5:1, you have to find out whether the antenna is too long or too short and act accordingly. If the SWR on Channel 1 is higher than on Channel 23, the antenna is too short. Properly tuned antennas are very low SWR on Channel 12 or 13 and equal SWR on Channel 1 and 23. If the SWR cannot be brought below 1.5:1, recheck the installation. You will find that most antennas are supplied proper length or overlength. Those antennas (fiberglass and steel whips), with no resonator tips can also be tuned by placing another lockwasher at the base to move it up slightly, if necessary.

If it is determined that the antenna is too long and must be cut, do it judiciously, for it is possible to go too far. Cut, file, or grind in about 1/16-1/8 inch increments, checking the SWR at Channel 1 and 23 each time until you are satisfied. When you are through the resonator tip should be fully inserted into the loading coil and the set screw tightened securely. Twin antennas must be trimmed IDENTICALLY.

THIEVERY

CB radio rip-off is becoming so popular that some motels advise customers to take their CBs inside. There is no surefire way to prevent someone from stealing your rig, unless you take it with you each and every time you leave your car. This gets to be monotonous, unless you have had the foresight to install your set with a slide mount, or to conceal it in some other trick installation.

There are several products on the market which will at least deter, if not thwart completely, most thieves. Since thieves must work speedily, something which will make them work harder is

A CB radio lockmount may deter a thief

sometimes a deterrent. There are locked barrels which completely cover the mounting nuts holding the radio in the bracket, locking brackets, and other similar methods. But thieves generally pry the mounting bracket loose and take the radio, locking bracket, and lock. This is one of the joys of an in-dash mount.

With this in mind, there are several steps to take:

1. Don't park in the evil parts of town and leave your radio in the vehicle.

2. When you do park on the streets or in a lot for short periods of time, try to park under a light.

3. When you are going to be gone for a while, remove the set and stow it out of sight. If you don't remove it, AT LEAST cover it with 'something.

4. Don't forget to remove the antenna (if you've equipped it with a quick-disconnect). If you don't, the thing sticks out like a divining rod.

5. If your set is installed in a small boat, take it with you when you leave the boat at night. It doesn't take that long and you don't do it that often.

6. Many local law enforcement agencies provide a number etch-

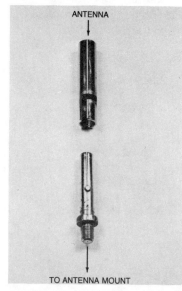

"Quick-disconnect" allow instant removal of the antenna

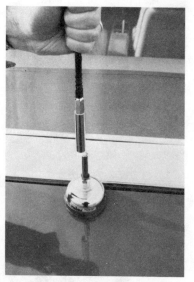

"Quick-disconnect" installed on a trunk mount. Take the antenna off if you're going to be gone long.

ing service. A number (usually social security) is etched onto your set and logged in police files. It won't keep the set from being stolen, but it may aid recovery.

INSTALLING BASE CBs

Installing a base set is not as complicated as a mobile rig. Very little is actually involved, other than installing the antenna, if you're using an outside antenna. Pick a suitable place to put the set, connect the power and antenna leads, and that's it.

Like mobile rigs, each antenna installation is different, but the antenna should be placed as high as possible; however, an excessively long cable may result in signal loss. Try to put the set as close as possible to the place where the cable enters the room to help minimize this, yet in a place where it will get ventilation.

In most instances coaxial cable is bought by the foot for base installation. It must be cut to length and the ends terminated with

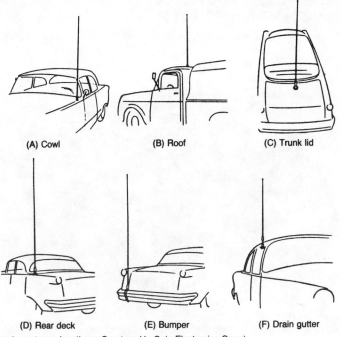

(A) Cowl   (B) Roof   (C) Trunk lid

(D) Rear deck   (E) Bumper   (F) Drain gutter

Popular antenna locations (Courtesy Hy-Gain Electronics Corp.)

Install a base station antenna as high as possible, within the FCC rules (Courtesy New Tronics)

AC to DC converters are available to use a mobile set as a temporary base (Courtesy Breaker Corp.)

the proper hardware. The antenna should be run through an access hole in the wall, which is then packed and sealed. Cables running between the sill and windows are for temporary installations only.

Consult the owner's manual for instructions regarding grounding. Sometimes a standard 5 ft. long copper ground rod, driven into the earth, is recommended for connection to the specified radio terminal with 12 gauge (or heavier) wire. Clamping the ground wire directly to a copper cold water pipe can also make an effective ground. Devices are also available to protect against lightning when using an ungrounded antenna, which are basically lightning and static electricity arrestors placed in series with the standard antenna cable connectors.

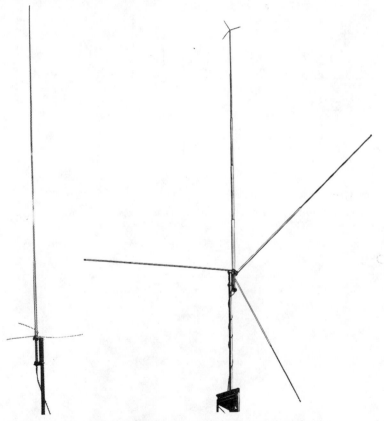

(Right) A ⅝ wave ground plane base station antenna (Courtesy Breaker Corp.)

(Left) Omnidirectional base station antenna (Courtesy The Antenna Specialists Co.)

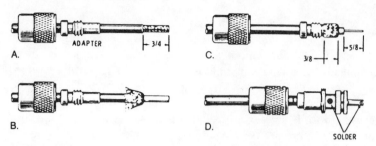

Installing a PL-259 connector on the coax (Courtesy The Antenna Specialists Co.)

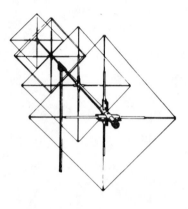

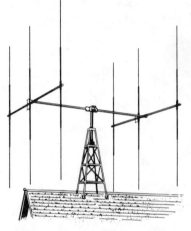

A "quad" base station antenna (Courtesy The Antenna Specialists Co.)

A directional base antenna (Courtesy The Antenna Specialists Co.)

Lightning arrestor

NOISE SUPPRESSION

Noise, or interference, is classified as one of two types—RFI or TVI. RFI (radio frequency interference), is interference with the ability of your CB set to receive signals and is commonly known as static. TVI (television interference), is generated by your CB set and will sometimes affect the TV, radio, or hi-fi radio reception in close proximity to your transmitter. Both are annoying depending on whether you're listening to your CB or your neighbors are trying to use their TV or hi-fi. The degree of noise suppression needed will vary in each different case. Each case is unique, requiring its own solution.

RFI NOISE

RFI noise falls into two classes, either man-made or natural. Natural noise, atmospheric disturbances, sunspots, lightning, etc., are a problem with CB, but little can be done about controlling them. Man-made noises produced by electric motors, vehicle electrical systems, farm machinery, medical equipment, and machinery of all types, are the main cause of noise, and, fortunately, you can do something about these.

Because of the low power used for CB, its phenomenal growth in the last few years, high receiver sensitivities, and the proximity to sources of interference, the RFI problem can be severe even though almost all CBs employ noise limiters. Vehicles built in the United States are provided at the factory with adequate suppression for reception on AM/FM receivers which conforms to SAE suppression standards, but CB, and all other types of two-way communication equipment, requires the same basic suppression techniques, but far more extensively and carefully.

In vehicles such as motorcycles, cars, trucks, boats, and RVs, noise can get into the transceiver in one or more of three ways: through the power source, through the antenna, or picked up by the internal circuitry of the receiver. Don't confuse background noise with interference. Concern yourself with it only if it disrupts reception at normal volume.

There are two fundamental approaches used to suppress noise: reduce the strength of the interference at the source; or, to confine the interference, using the engine compartment as a shielding box. Capacitors, bonding, routing of wiring, and high-voltage suppressors are the basic hardware and techniques used.

CAPACITORS

A capacitor is designed to pass the flow of alternating current, but to block the flow of direct current. Interference of this type (man-made), is almost always an alternating or impulse type of signal and the capacitor will direct most of the flow of this type of current to ground without affecting the circuit of the direct current. A conventional by-pass capacitor is suitable for the broadcast band, but for effective suppression with higher UHF frequencies, the use of coaxial capacitors is recommended.

CAUTION: *Capacitors should never be used on transistorized or electronic ignitions.*

CONVENTIONAL BYPASS CAPACITOR

COAXIAL CAPACITOR

Two types of capacitors (Courtesy Champion Spark Plug Co.)

BONDING

Bonding is a particular technique used to connect the metal parts of the vehicle together to form an effective shield blocking RFI. Interference generated by the ignition and charging systems will be kept from traveling throughout the vehicle and a common ground will be formed for all RFI signals.

WIRE ROUTING

Wire routing must be carefully done; if not, interference will be transferred from one circuit to another, particularly to the high-voltage, or ignition cable side.

HIGH-VOLTAGE SUPPRESSORS

The ignition system is probably the greatest single source of RFI in a vehicle, and resistors are available to reduce the interference to a tolerable level.

GENERAL SUPPRESSION PROCEDURES

Anytime two-way radio or audio equipment is replaced or serviced, the following steps will help minimize the need for additional suppression.

NOTE: *Before attempting any of the procedures outlined in this section, disconnect the wires from the battery. If you don't, you could be seriously injured.*

1. Be sure that all of the original equipment for suppression is still intact and in good condition. It's possible that resistor cable could have been replaced by non-resistor cables, a bonding strap could have been removed, or a toothed lockwasher may have been lost.

2. Be sure that all components and connections are in good condition. A corroded connection will, in all likelihood, make interference worse.

3. Tune the engine or have it tuned by a specialist. Tune-up

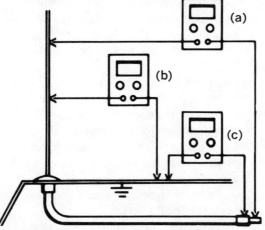

3 checks you can make on your antenna with an ohmmeter (Courtesy Champion Spark Plug Co.)

should include new spark plugs, points, and condenser at the least. Additional items which should be looked at, but require replacement less frequently, are the cap and rotor. Optimum radio performance will not be delivered unless the ignition system is in good condition.

4. Ideally, the radio should be connected to the battery. Connecting it to the accessory or ignition side of the ignition switch leads to interference in the radio from the car's electrical system.

5. Low-voltage wires should be kept away from the ignition system, as well as any other circuits which are suspected noise producers. Wires of suspected circuits should be laid flat against a grounded metal area where possible; they should not be bundled together.

6. Be sure that the antenna lead-in shield is grounded at both ends. Insulation, as well as all connections, should be clean and tight.

If you possess an ohmmeter, there are three checks you can make on the antenna. If you don't own an ohmmeter, try to borrow one to make these three checks.

a. Put the ohmmeter on the lowest scale and touch the prods to the antenna rod and to the center contact of the plug. The resistance should be a fraction of an ohm;

b. Put the ohmmeter on the highest scale. Touch the ohmmeter prods to the antenna rod and to the vehicle ground. Most antennas should read an open circuit, except for the few high "gain" type antennas with built-in transformers, which will be short-circuited;

c. Return the ohmmeter to the lowest scale. Touch the prods to the outside of the antenna plug and to the vehicle ground. Resistance should be zero.

If any of these tests don't turn out as they should, there is a serious fault or open circuit in the antenna system.

7. Above all, good suppression can only take place if all components are properly connected and grounded. All paint, oil, grease, or rust should be removed from all areas where good electrical contact is essential. Clean hardware and sharp-toothed washers should be used for mounting components. All places where lugs or eyes have been attached should be soldered to the wire, and all electrical connections should be taped.

If, after you have done all this and the evil noises still persist, you will have to conduct a step-by-step search to identify the culprit.

IDENTIFYING INTERFERENCE

Each type of interefernce you hear on the receiver has its own distinctive sound and characteristics. In order to find out what is causing the interference, you at least have to know where to start looking.

IGNITION SYSTEM: This is a popping sound which increases in tempo with the engine speed. It will also shut off immediately when the ignition key is turned off at fast idle.

GENERATOR/ALTERNATOR: These produce a musical whine, high-pitched, increasing in frequency with higher engine speed. It will not shut off instantly when the ignition key is turned off at fast idle.

VOLTAGE REGULATOR: Voltage regulator interference is usually heard in conjunction with alternator or generator noise, and makes its appearance as a rasping, ragged sound occurring at an irregular rate. It will not stop instantly when the ignition is shut off at fast idle.

INSTRUMENTS: Instruments in the dash produce hissing, crackling, and clicking sounds occurring at irregular intervals as the gauges operate. The condition is usually worse on rough roads and can be tested by jarring the dashboard.

The voltage limiter behind the dash, which is used with the fuel and temperature gauges, can produce a loud "hashing" sound at intermittent intervals. Bouncing the vehicle to activate the fuel gauge sending unit should verify RFI from the voltage limiter.

Disconnect the gauges or the sending units one at a time; the RFI should disappear if they are at fault.

ACCESSORIES: Make a preliminary check with all accessories turned off. Turn them on one at a time and listen for increased RFI. Intermittent noise from the turn signal or hazard

SMALL CLIP (HOT)   CAPACITOR

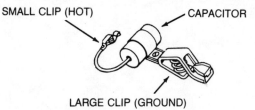

LARGE CLIP (GROUND)

Using a grounded capacitor to identify the source of interference by the process of elimination (Courtesy Champion Spark Plug Co.)

warning flashers or windshield wipers can often be eliminated by the use of a capacitor, but most by-pass capacitors will have no effect on wiper motor noise. Coaxial capacitors should be used for this.

WHEELS & TIRES: Wheels and tires sometimes create a pop-ping or rushing sound through the radio while they operate on dry roads at high-speeds. Interference from the wheels and tires can be traced by lightly applying the brakes; the noise should disappear.

OTHER SOURCES: If a particular type of interference cannot be identified as coming from any of the sources described above, a test capacitor can be easily constructed as shown. A grounded capacitor touched to all "hot" electrical connections will identify the offending item if the RFI disappears.

Another test instrument can be constructed at home which is very useful in locating the source of RFI. Begin by disconnecting the antenna from the receiver. Wrap 50 turns of insulated, or bell, wire into a coil 2 inches in diameter, and tape the coil of wire to a broom stick or wooden dowel rod as shown. Using a few feet of normal lamp cord, connect one side of the coil to an alligator clip which will be used for the ground side. The other end of the coil should be connected to the center conductor of a PL-259 connector which can be purchased from any electronics or CB store. Basi-cally, what you have done is to construct a crude inductive antenna which will pick up interference. Connect the PL-259 to your radio, start the engine, and turn the radio on. Probe around the engine and wiring with your homemade coil. Interference will be the loudest when you are close to the source of the interference.

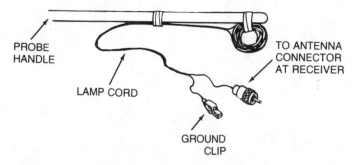

PROBE
HANDLE

TO ANTENNA
CONNECTOR
AT RECEIVER

LAMP CORD

GROUND
CLIP

Homemade antenna for locating the source of interference by probing (Courtesy Champion Spark Plug Co.)

RFI SUPPRESSION TECHNIQUES

CAUTION: *Capacitors should not be used on transistorized or electronic ignitions.*

ALTERNATOR

The alternator slip-rings should be clean and the brushes should make good contact. A 0.5 mfd (microfarad), coaxial capacitor can be installed at the alternator output terminal. Be sure that it is rated to handle the alternator output current.

NOTE: *Do not connect a capacitor to the alternator field terminal.*

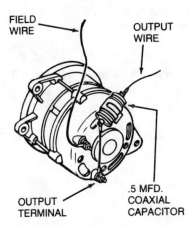 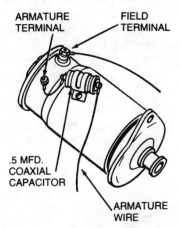

Capacitor installed at the alternator output terminal (Courtesy Champion Spark Plug Co.)

Capacitor installed at the generator output terminal (Courtesy Champion Spark Plug Co.)

GENERATOR

Most American (or import), cars and trucks these days are not equipped with generators. But, for the cars which are, the commutator and brushes should be making good contact. If the commutator is badly worn, the generator should be overhauled.

Remove the factory-installed capacitor from the armature terminal and install a 0.5 mfd coaxial capacitor which is rated to handle the current output of the generator.

NOTE: *Do not connect the capacitor to the generator field terminal.*

VOLTAGE REGULATOR

Many of the newer cars are now using solid state regulators, often built into the alternator. But, for those cars still equipped with the traditional external voltage regulator of the single or double-contact type, a 0.5 mfd capacitor can be installed as close as possible to the *armature* and *battery* terminals. On a single-contact regulator, use a 0.5 mfd capacitor at the ignition terminal. Again, be sure that the capacitor(s) are rated to handle the generator or alternator current output. The rated output can be found in the electrical specifications of most any service manual for your car.

Capacitors should not be connected to the regulator *field* terminal. Unusual cases of interference may require that the FIELD wire be shielded. In this case, be sure that both ends of the shield are grounded.

If regulator noise is extreme or simply cannot be quieted, the wire from the "F" or field terminal can be replaced with a piece of RG-8/U coaxial cable. If you do this, be sure that the coaxial cable does not touch the engine block or any other accessory delivering a lot of heat. Also, be sure that the braid at the ends of the coaxial cable is securely grounded to the chassis or nearest ground point other than the engine.

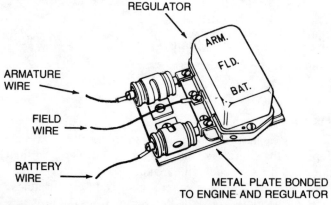

Capacitors installed on the voltage regulator (Courtesy Champion Plug Co.)

DC POWER LINES

The DC power line to your set is a potential source of a great deal of noise. A capacitor can be installed in this line by mounting it on

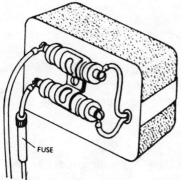

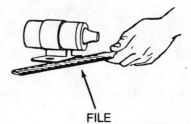

FILE

Clean the back of the coil mounting bracket
(Courtesy Champion Spark Plug Co.)

FUSE

Capacitors installed on a DC power line (Courtesy Sprague Electronics)

the radio cabinet. Cut into the power line as close as possible to the
radio.

Another trick used by ham operators, but applicable to CBs is to
use coaxial antenna cable for the power line. The center conductor
of the coaxial cable is used for the "hot" lead connection after the
inline fuse, and the coaxial shielding is used for the ground connection.

INSTRUMENTS

A 0.5 mfd capacitor installed at the terminals of the gauges or
sending units will usually silence interference from these sources.

The voltage limiter can usually be quieted with a 0.5 mfd
capacitor connected at the battery terminal of the voltage limiter.
In place of this, a 0.1 mfd radio-type pigtail capacitor connected
across the voltage limiter terminals also will work. Extreme cases

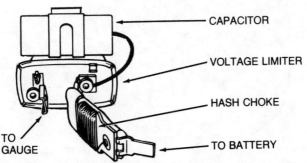

CAPACITOR

VOLTAGE LIMITER

HASH CHOKE

TO
GAUGE

TO BATTERY

Hash choke installed in series at the voltage limiter (Courtesy Champion Spark Plug Co.)

of noise from the voltage limiter can be cured by installing a ''hash choke'' in series with the battery lead.

ACCESSORIES

Almost any accessory which is operated by a brush motor (turn signals, stop signals, electric windows, heater blowers, and the like), can be quieted with a 0.25 mfd capacitor installed at the accessory terminals.

BONDING

Bonding straps can be pieces of 1/2-1 inch wide copper braided strap for connecting components to ground, or pieces of metal for grounding fenders. Braided copper straps can be obtained from

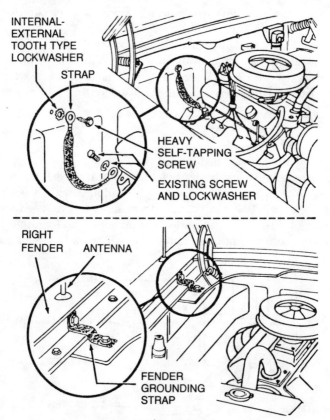

INTERNAL-EXTERNAL TOOTH TYPE LOCKWASHER

STRAP

HEAVY SELF-TAPPING SCREW

EXISTING SCREW AND LOCKWASHER

RIGHT FENDER   ANTENNA

FENDER GROUNDING STRAP

Examples of bonding (Courtesy Champion Spark Plug Co.)

most well-stocked electronic supply stores. In addition, many car manfacturers and radio manufacturers offer bonding kits of this material. If you can't find the copper stuff, an alternative is to use the braided coaxial shielding from a piece of spare coaxial antenna cable. You can get this off by carefully slitting the outside insulation from the coaxial cable (without slitting the shielding). Peel away the insulation and slip the braided shielding off the foam insulation. Whatever type of bonding material is used, be sure that the lugs used to attach the cable are securely attached to the cable and soldered.

The art of "bonding" is largely a matter of luck and trial-and-error. The location of the bonding straps often plays an important role in its effectiveness, and experience will most times reveal the best location. An expert at CB installations can offer words of wisdom on this subject.

Some good places to begin installing bond straps are:

Corners of the engine to the frame;
Exhaust pipe to the frame and engine;
Both sides of the trunk and hood lids;
Coil and distributor-to-engine and firewall;
Air cleaner-to-engine;
Battery ground-to-frame;
Tailpipe-to-frame;
Steering column, oil pressure gauge line and any other
    metal lines passing through the firewall;
Front and rear bumper supports; and
Radiator-to-radiator supports.

Generally, any metal parts which are separated from the frame by any type of insulation (spacers, paint, noise silencing material), should be electrically connected to the frame, or connected together.

Use self-tapping screws in conjunction with toothed lockwashers to cut into surface layers of metal. Bonding straps should be as short and heavy as possible to be really effective, and should be checked periodically for corrosion and tightness.

WHEELS

Static collector rings, installed inside the front wheel caps, will collect static build-up from the front wheels and prevent it from entering the receiver.

PRIMARY IGNITION SYSTEM

IGNITION COIL

The first step is to remove the ignition coil and its mounting bracket. Clean the paint from the back of the bracket with sandpaper or a file and from the mounting point on the engine. Reassemble the bracket and the ignition coil tightly.

If this does not help, install a 0.1 mfd coaxial capacitor as close to the coil battery terminal as possible. Do not connect the coaxial capacitor to the distributor terminal, nor should a normal by-pass capacitor be used. In addition, a 0.005 mfd 1,000 volt ceramic disc capacitor installed at the coil distributor terminal will help eliminate interference.

Be sure to check coil polarity, or have it checked.

DISTRIBUTOR BREAKER POINTS

The distributor breaker points are not usually the cause of interference, although they cannot be totally ignored as a cause of RFI. Point condition determines to a large extent whether or not they

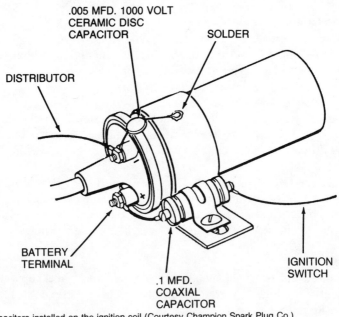

Capacitors installed on the ignition coil (Courtesy Champion Spark Plug Co.)

will produce any interference. Points which have been well-maintained are far less likely to cause any interference problems than are those which have been poorly maintained. If the points are suspected of causing a great deal of interference, look for point bounce or deteriorated points.

Fleets and commercial trucks can reduce RFI and at the same time increase point life by determining the allowable condenser limits for the vehicle, and selecting condensers at the high limit for mainly stop/start driving or selecting condensers at the low limit for mainly high-speed driving. If the condenser limits are determined to be 0.18-0.25 mfd, use a 0.25 mfd condenser for stop/start driving or a 0.18 mfd condenser for high-speed use.

NOTE: *By-pass capacitors should not be installed at the coil distributor terminal.*

DISTRIBUTOR CAP AND ROTOR

The distributor cap and rotor should be replaced at the interval specified by the vehicle manufacturer. In terms of suppressing RFI, they should be replaced when the tip of the rotor and the contacts in the cap show signs of erosion or carbon tracking, which increases the gap over which the spark must jump, leading to higher RFI levels.

Replace the rotor and distributor cap when they show signs of wear to help reduce interference (Courtesy Champion Spark Plug Co.)

The rotor used with a GM V-8 distributor can be had in a radio suppression version, which is stamped with an "E" on the metal blade.

SECONDARY IGNITION SYSTEM

This is the high-voltage side of the ignition system and the worst producer of RFI. Radiated interference is reduced effectively by suppressor resistors of various types. Some are separate components, while others have been incorporated into the distributor rotor or spark plug towers on the cap. These are mainly service items for U.S. cars and trucks.

SPARK PLUG CABLES

SAE standards specify two resistance ranges for cable use on newly-manufactured vehicles:

LR—3,000-7,000 ohms per foot;

HR—6,000-12,000 ohms per foot.

Of the two, LR is the most common, but HR is sometimes used between the coil and distributor where short cables are involved.

Replacement wires of the resistance type are available from any

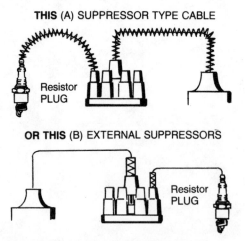

THIS (A) SUPPRESSOR TYPE CABLE

Resistor PLUG

OR THIS (B) EXTERNAL SUPPRESSORS

Resistor PLUG

Do not combine spark plug suppression devices (Courtesy Champion Spark Plug Co.)

of several manufacturers. The resistance data is generally available on the box or from the manufacturer.

When handling suppressor cables, never pull on the cable; remove them by pulling on the boot. Never try to attach a screw-on suppressor to a suppressor cable and don't try to repair suppressor cables. Cables which are damaged should be replaced.

RESISTOR SPARK PLUGS

Spark plugs of the type known as "resistors" afford better protection from RFI than conventional spark plugs. The use of resistor-type spark plugs is increasing in newer engines because of their ability to maintain their suppressive characteristics for long periods of time.

Spark plugs should also be maintained properly to provide protection against RFI. A wide plug gap, when the electrodes are burnt, or uneven, requires higher than normal voltage causing the ignition system to emit higher than normal levels of RFI.

Combining Suppression Devices

Some cars and trucks are equipped with resistor-type plugs and suppressor spark plug cables. Resistor-type plugs can be used in other vehicles where more suppression is desired, or some other suppression devices can be used, but combining the two is not a good idea. Choose between:

Resistor plugs with suppressor cables,
OR
Resistor plugs with 10,000 ohm suppressors in the center tower of the distributor and 5,000 ohm suppressors in the spark plug distributor towers.

SAFETY WARNING DEVICES

Federal safety regulations have required the addition of a variety of warning lights and buzzers over the past few years. These buzzers are to remind the driver to turn the headlights off, fasten seatbelts, remove the ignition key, etc. If necessary, these circuits can be suppressed at the primary power circuits.

Be careful when hooking up your transceiver to a power source—it is not a good idea to connect it to any source shared with a buzzer, solenoid, or flasher.

INTERFERENCE RECEIVED ON BASE SETS

Any electrical device which uses moving contacts, such as TVs fluorescent lamps, sewing machines, thermostats, and the like, are likely to produce some interference in a base set receiver.

To avoid this, place the receiver as far as possible from appliances, particularly TV sets or fluorescent lights. If the interference is particularly bad from some appliances, an AC power line filter can be installed between the AC outlet and the appliance. This will be particularly effective if the appliance power line is radiating interference.

Spurious signals can also be generated and mixed with other spurious signals. If you happen to be turned to the middle of these spurious signals, it is possible that you could hear both on your receiver. This will generally occur only with antennas close to one another. The remedy is to get a greater separation between the antennas, possibly by mounting one higher than the other.

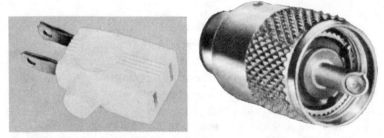

AC power line filter   (Courtesy GC Electronics)   CB antenna dummy load

CB RADIO FOR MARINE USE

If you are using your CB on a boat, it is probably very susceptible to RFI noise. Automotive type suppressive measures are used where applicable, but special problems exist which require special handling.

OUTBOARD MOTORS

Outboard motor manufacturers offer suppression kits for their engines. The hardware in these kits should cure most of the RFI problem on an outboard if the directions with the kit are followed

Outboard motors which tend to foul spark plugs quickly may do so even quicker if suppresive equipment is installed. Before installing any suppressors, find and correct the cause of the fouling. It is even possible that outboard motor spark plugs will become more susceptible to carbon build-up between the center and outside electrodes during prolonged operation at low speeds. Because of the increased chances of this with radio suppression equipment installed, it is a good idea to check the plugs at more frequent intervals than before.

If in doubt about radio suppression on an outboard, check with your dealer.

INBOARD ENGINES

Inboard engines installed in metallic hulls have the built-in metal shielding, similar to the body of an automobile, and are, therefore, more adequately grounded than engines in wooden or fiberglass hulls. Basic automotive suppression techniques apply here also, but, again, special additional techniques are necessary.

Most inboard boat owners have copper ground plates attached to the hull. A great deal of the interference generated by the engine can be suppressed by shielding the entire engine compartment with copper or bronze screening connected to the ground plate. In addition, kits to shield the engine compartment are commercially aviable and easily installed.

REMOTE CONTROL PANEL

The remote control panel for inboard engines is a source of high level RFI.

The interference level can be greatly reduced by shielding the entire length of the tachometer wire from the control panel to the engine and grounding the shield at both ends. Before doing this, however, check with the manufacturer of the tach or a marina—some tachometers will not work if the shielding is added.

The ignition switch wires from the engine to the control panel can likewise be shielded. Shielding should be installed over the entire cable assembly and grounded at both ends.

TIPS FOR GOOD MARINE RECEPTION

To obtain optimum marine reception, DO the following:
1. Use an approved marine-type antenna.

2. Install the antenna on the opposite side of the boat from the remote control panel.

3. Terminate all ground leads from electronic equipment at the negative side of the battery, not the ground plate. The ground plate should be grounded to the negative side of the battery with a heavy copper strap to minimize the effects of electrolysis. Wood or fiberglass boats should use commercially available radio ground plates or at least a 12 sq. foot underhull ground plate.

4. If possible, all wiring should be routed under the rails of a metal boat. Metal boats do not require ground plates if the electrical system is grounded to the hull.

5. Eliminate interference from accessories with 0.25 mfd capacitors at the terminals of such equipment as bilge pumps, windshield wipers, etc.

6. Check the seldom-recognized sources of interference such as metal rigging rubbing together, loose metal fittings giving off vibrations at certain propeller frequencies and dissimilar underwater metal fastenings generating voltage.

TELEVISION INTERFERENCE (TVI)

TVI, or television interference, applies mostly to base sets, whose operators, like ham operators, are roundly criticized as the cause of interference picked up by neighboring television sets, hi-fi sets, and radios. If you operate a base station, it'll pay to know something about TVI, if only to preserve peace and harmony in the neighborhood.

It is just as true that a transmitter operated in close proximity to a TV set can cause interference on the TV, as it is that the CB set may not be at fault. If only a few VHF channels pick up TVI, then the transmitter is a likely suspect, but, if TVI is noticed on all channels, the TV is the place to start looking. The reason for TVI lies in two facts. First, under ideal conditions, an AM CB transmitter should put out a carrier signal plus upper and lower sidebands only, and an SSB transmitter should generate only the carrier and/or one sideband. Second, ideal conditions seldom exist. Signals known as "spurious signals" are also generated. The strongest spurious signal is the second harmonic of the generated signal. Briefly, the second harmonic of a given frequency is twice the frequency (i.e., the second harmonic of CB Channel 10 at 27.075 MHz is 54.15 MHz).

Prior to 1974, FCC type acceptance was not required for any CB set sold. However, all sets sold after 1974 must be FCC type accepted. One of the criteria for type acceptance is that spurious signals must be attenuated by at least 50 db, meaning that on a full 4 watt output, the spurious signal generated cannot be more than 40 microwatts in power, which is not sufficient to cause TVI.

ELIMINATING TVI

To deal effectively with the problem of TVI, it must first be determined whether the TV set or CB transmitter is at fault. Begin by making the following quick test.

1. Set the TV to each of the VHF channels.

2. At each channel, in turn, key the mike on your CB set. On an AM set, it is only necessary to key the mike; on an SSB transmitter, you will have to speak into the mike, since no signal will be transmitted unless modulation takes place.

3. Look at the following chart. If TVI is present on more than one harmonically-related channel, the transmitter may be at fault.

TRANSMITTER HARMONICS AFFECTING TV CHANNELS
(Numbers under frequency band indicate harmonic signal)

TV Channel	CB Frequency 27 MHz
2	2nd
3	—
4	—
5	—
6	3rd
7	—
8	—
9	7th
10	7th
11	7th
12	7th
13	—

Because the 2nd harmonic is the strongest, TV Channel 2 is particularly susceptible to TVI.

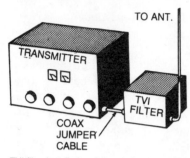

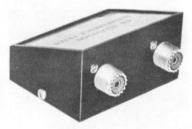

Typical TVI filter (Courtesy GC Electronics)

TVI filter installed on base station transmitter

Obviously other harmonics of different frequencies will also affect TV reception on some channels, but these apply only to CB frequency.

If it is determined that the transmitter is possibly at fault, the next step is to replace the antenna with a dummy load, available at all electronic or CB stores. If the TVI persists, even with the dummy load in place of the antenna, you can be pretty sure that the harmonic frequency is being given off by the power line or by the transmitter. This is known as harmonic attenuation, and a power line filter installed at the transmitter will probably help if the signal is being radiated from the power line. If the signal is being given off

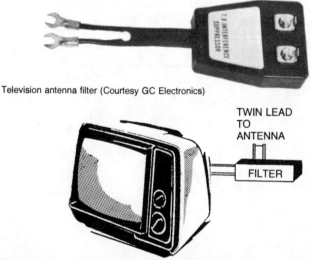

Television antenna filter (Courtesy GC Electronics)

Television antenna filter connects between the TV and the two antenna leads by the transmitter, an inexpensive low-pass filter should be installed at the transmitter. A few sets are even equipped with a TVI adjustment.

If the TV receiver is determined to be at fault (TVI will be present on most or all of the VHF channels), the TV antenna or the transmission line is picking up the interference, known as "swamping" the input side of the TV. An inexpensive high-pass filter, connected at the terminals of the TV set, will alleviate this condition.

AUDIO RECTIFICATION INTERFERENCE

Occasionally, CBers receive complaints that their voices are being heard through a TV set, radio, hi-fi, or tape deck. This is what engineers call audio rectification interference. The audio portion of the hi-fi, or whatever, is picking up your signal and reproducing it through the speakers. This is not the fault of the CB transmitter. It is possible that in the design and manufacture of the audio equipment involved, this problem was not foreseen, and the shielding or circuit design is inadequate to cope with the problem or the radio signal could also be picked up by audio cables, amplifier wiring, or speaker cables.

Interference of this type on a TV set can be controlled by installing a high-pass filter at the TV antenna filter. If this doesn't work, a qualified technician should apply filtering at the TV audio amplifier.

In hi-fi systems, a low-pass filter connected in series with each audio input cable may do the job. If the speaker cables are picking up the interference, a 0.01 mfd disc capacitor should be soldered across each of the amplifier outputs. Extreme cases will probably require the services of an electronic technician to install filters in the amplifier itself.

The power cable of the hi-fi or TV set may, on occasion, be at fault. In these cases, a power line filter may suppress the interference. Most of these filters are intended for use only in properly grounded electrical systems, such as a 3-wire 110 volt AC system or a permanent chassis-to-ground system. Before performing any work on AC line-operated equipment, pull the plug!

Remove the line cord from the AC receptacle. Any filters or by-pass capacitors connected across the line should be discharged by grounding them. Disconnect the line cord from the terminal strip, fuse holder, or transformer, and remove any by-pass (ceramic disc) capacitors. Install the AC line filter so that its lead wires reach the point where the line cord was originally connected. Attach the filter, making sure that a good connection is obtained, and connect the lead wires where the AC line cord was previously connected. Connect the original AC line cord to the filter as shown.

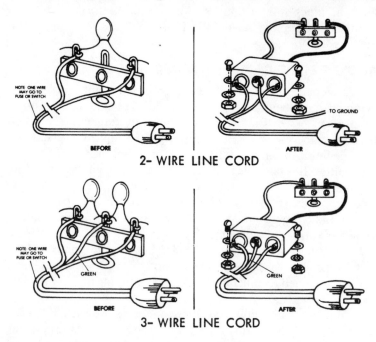

AC power line filter installed (Courtesy Sprague Electronics)

FCC RULES AND REGULATIONS

The Federal Communications Commission Rules and Regulations, Part 95 are the bible as far as CB radio is concerned. It governs everything which has to do with the Citizens Radio Service, or CB to you and I. When you sign your application for a station license, you aver that you have or have ordered from the Federal Government, Part 95 of the Commission's rules governing the Citizens Radio Service. But, just to keep you informed and provide some light reading while you wait for your very own copy to arrive (it will take a while), we've reprinted a copy here.

As of September 1, 1975, there are a few changes to Part 95 which have not yet been printed.

First, and, most importantly, the "hobby" restriction has been done away with forever. This means that the FCC will no longer limit what you say on the air, except in cases of profanity, playing of music, or selling of merchandise. This change was made to help eliminate the fear of identification and you will not be cited after September 1, 1975 for radio checks, local gossip, or discussions of other similar nature.

Second, Channel 11 is now the national calling channel. This channel is now to be used for calling and contact ONLY. Once contact is established, find another channel.

Other changes include:

There is no longer a distinction between intra- and interstation calls. Any licensee can call any other.

It is now OK to use your "handle," once you have identified yourself by your station call letters. Call letters should still be used at the beginning and end of conversations, but in between, "handles" are alright.

The 5 minute limit on transmissions is still in effect, but the waiting time between transmissions has been reduced from 5 minutes to 1 minute.

It is now permissible to relay communications provided that they are relayed no farther than 150 miles from the original point of transmission.

Antenna height restrictions now apply to transmitting and receiving antennas. Previously, they applied to transmitting antennas only.

That's about it for rule changes, although these made will have a considerable impact, finally making CB a legal hobby.

The Federal Communications Commission Rules and Regulations Part 95 is for sale from the Superintendent of Documents, United States Government Printing Office, at $5.35 per copy.

CONTENTS—PART 95

95.97 Operator license requirements.
95.101 Posting station license and transmitter identification cards or plates.
95.103 Inspection of stations and station records.
95.105 Current copy of rules required.
95.107 Inspection and maintenance of tower marking and lighting, and associated control equipment.
95.111 Recording of tower light inspections.
95.113 Answers to notices of violations.
95.115 False signals.
95.117 Station location.
95.119 Control points, dispatch points, and remote control.
95.121 Civil defense communications.

Subpart E—Operation of Citizens Radio Stations in the United States by Canadians

95.131 Basis, purpose and scope.
95.133 Permit required.
95.135 Application for permit.
95.137 Issuance of permit.
95.139 Modification or cancellation of permit.
95.141 Possession of permit.
95.143 Knowledge of rules required.
95.145 Operating conditions.
95.147 Station identification.

AUTHORITY: §§ 95.1 to 95.147 issued under secs. 4, 303, 48 Stat. 1066, 1082, as amended; 47 U.S.C. 154, 303. Interpret or apply 48 Stat. 1064–1068, 1081–1105, as amended; 47 U.S.C. Sub-chap. I, III–VI.

SUBPART A—GENERAL

§ 95.1 Basis and purpose.

The rules and regulations set forth in this part are issued pursuant to the provisions of Title III of the Communications Act of 1934, as amended, which vests authority in the Federal Communications Commission to regulate radio transmissions and to issue licenses for radio stations. These rules are designed to provide for private short-distance radiocommunications service for the business or personal activities of licensees, for radio signaling, for the control of remote objects or devices by means of radio; all to the extent that these uses are not specifically prohibited in this part. They also provide for procedures whereby manufacturers of radio equipment to be used or operated in the Citizens Radio Service may obtain type acceptance and/or type approval of such equipment as may be appropriate.

§ 95.3 Definitions.

For the purpose of this part, the following definitions shall be applicable. For other definitions, refer to Part 2 of this chapter.

(a) Definitions of services.

Citizens Radio Service. A radiocommunications service of fixed, land, and mobile stations intended for short-distance personal or business radiocommunications, radio signaling, and control of remote objects or devices by radio; all to the extent that these uses are not specifically prohibited in this part.

Fixed service. A service of radiocommunication between specified fixed points.

Mobile service. A service of radiocommunication between mobile and land stations or between mobile stations.

(b) Definitions of stations.

Base station. A land station in the land mobile service carrying on a service with land mobile stations.

Class A station. A station in the Citizens Radio Service licensed to be operated on an assigned frequency in the 460–470 MHz band with a transmitter output power of not more than 50 watts.

Class B station. (All operations terminated as of November 1, 1971.)

Class C station. A station in the Citizens Radio Service licensed to be operated on an authorized frequency in the 26.96–27.23 MHz band, or on the frequency 27.255 MHz, for the control of remote objects or devices by radio, or for the remote actuation of devices which are used solely as a means of attracting attention, or on an authorized frequency in the 72–76 MHz band for the radio control of models used for hobby purposes only.

Class D station. A station in the Citizens Radio Service licensed to be operated for radiotelephony, only, on an authorized frequency in the 26.96–27.23 MHz band and on the frequency 27.255 MHz.

Fixed station. A station in the fixed service.

Land station. A station in the mobile service not intended for operation while in motion. (Of the various types of land stations, only the base station is pertinent to this part.)

Mobile station. A station in the mobile service intended to be used while in motion or during halts at unspecified points. (For the purposes of this part, the term includes hand-carried and pack-carried units.)

(c) Miscellaneous definitions.

Antenna structures. The term "antenna structures" includes the radiating system, its supporting structures and any appurtenances mounted thereon.

Assigned frequency. The frequency appearing on a station authorization from which the carrier frequency may deviate by an amount not to exceed that permitted by the frequency tolerance.

Authorized bandwidth. The maximum permissible bandwidth for the particular emission used. This shall be the occupied bandwidth or necessary bandwidth, whichever is greater.

Carrier power. The average power at the output terminals of a transmitter (other than a transmitter having a suppressed, reduced or controlled carrier) during one radio frequency cycle under conditions of no modulation.

Control point. A control point is an operating position which is under the control and supervision of the licensee, at which a person immediately responsible for the proper operation of the transmitter is stationed, and at which adequate means are available to aurally monitor all transmissions and to render the transmitter inoperative.

Dispatch point. A dispatch point is any position from which messages may be transmitted under the supervision of the person at a control point.

Double sideband emission. An emission in which both upper and lower sidebands resulting from the modulation of a particular carrier are transmitted. The carrier, or a portion thereof, also may be present in the emission.

External radio frequency power amplifiers. As defined in § 2.815(a) and as used in this part, an external radio frequency power amplifier is any device which, (1) when used in conjunction with a radio transmitter as a signal source is capable of amplification of that signal, and (2) is not an integral part of a radio transmitter as manufactured.

Harmful interference. Any emission, radiation or induction which endangers the functioning of a radionavigation service or other safety service or seriously degrades, obstructs or repeatedly interrupts a radiocommunication service operating in accordance with applicable laws, treaties, and regulations.

Man-made structure. Any construction other than a tower, mast or pole.

Mean power. The power at the output terminals of a transmitter during normal operation, averaged over a time sufficiently long compared with the period of the lowest frequency encountered in the modulation. A time of $^1/_{10}$ second during which the mean power is greatest will be selected normally.

Necessary bandwidth. For a given class of emission, the minimum value of the occupied bandwidth sufficient to ensure the transmission of information at the rate and with the quality required for the system employed, under specified conditions. Emissions useful for the good functioning of the receiving equipment, as

for example, the emission corresponding to the carrier of reduced carrier systems, shall be included in the necessary bandwidth.

Occupied bandwidth. The frequency bandwidth such that, below its lower and above its upper frequency limits, the mean powers radiated are each equal to 0.5% of the total mean power radiated by a given emission.

Omnidirectional antenna. An antenna designed so the maximum radiation in any horizontal direction is within 3 dB of the minimum radiation in any horizontal direction.

Peak envelope power. The average power at the output terminals of a transmitter during one radio frequency cycle at the highest crest of the modulation envelope, taken under conditions of normal operation.

Person. The term "person" includes an individual, partnership, association, joint-stock company, trust or corporation.

Remote control. The term "remote control" when applied to the use or operation of a citizens radio station means control of the transmitting equipment of that station from any place other than the location of the transmitting equipment, except that direct mechanical control or direct electrical control by wired connections of transmitting equipment from some other point on the same premises, craft or vehicle shall not be considered to be remote control.

Single sideband emission. An emission in which only one sideband is transmitted. The carrier, or a portion thereof, also may be present in the emission.

Station authorization. Any construction permit, license, or special temporary authorization issued by the Commission.

§ 95.5 Policy governing the assignment of frequencies.

(a) The frequencies which may be assigned to Class A stations in the Citizens Radio Service, and the frequencies which are available for use by Class C or Class D stations are listed in Subpart C of this part. Each frequency available for assignment to, or use by, stations in this service is available on a shared basis only, and will not be assigned for the exclusive use of any one applicant; however, the use of a particular frequency may be restricted to (or in) one or more specified geographical areas.

(b) In no case will more than one frequency be assigned to Class A stations for the use of a single applicant in any given area until it has been demonstrated conclusively to the Commission that the assignment of an additional frequency is essential to the operation proposed.

(c) All applicants and licensees in this service shall cooperate in the selection and use of the frequencies assigned or authorized, in order to minimize interference and thereby obtain the most effective use of the authorized facilities.

(d) Simultaneous operation on more than one frequency in the 72–76 MHz band by a transmitter or transmitters of a single licensee is prohibited whenever such operation will cause harmful interference to the operation of other licensees in this service.

§ 95.6   Types of operation authorized.

(a) Class A stations may be authorized as mobile stations, as base stations, as fixed stations, or as base or fixed stations to be operated at unspecified or temporary locations.

(b) Class C and Class D stations are authorized as mobile stations only; however, they may be operated at fixed locations in accordance with other provisions of this part.

§ 95.7   General citizenship requirements.

A station license shall not be granted to or held by a foreign government or a representative thereof.

[*§ 95.7 revised eff. 2–5–75; VI(75)–1*]

SUBPART B—APPLICATIONS AND LICENSES

§ 95.11   Station authorization required.

No radio station shall be operated in the Citizens Radio Service except under and in accordance with an authorization granted by the Federal Communications Commission.

§ 95.13   Eligibility for station license.

(a) Subject to the general restrictions of § 95.7, any person is eligible to hold an authorization to operate a station in the Citizens Radio Service: *Provided,* That if an applicant for a Class A or Class D station authorization is an individual or partnership, such individual or each partner is eighteen or more years of age; or if an applicant for a Class C station authorization is an individual or partnership, such individual or each partner is twelve or more years of age. An unincorporated association, when licensed under the provisions of this paragraph, may upon specific prior approval of the Commission provide radiocommunications for its members.

NOTE: While the basis of eligibility in this service includes any state, territorial, or local governmental entity, or any agency operating by the authority of such governmental entity, including any duly authorized state, territorial, or local civil defense agency, it should be noted that the frequencies available to stations in this service are shared without distinction between all licensees and that no protection is afforded to the communications of any station in this service from interference which may be caused by the authorized operation of other licensed stations.

(b) [Reserved]

(c) No person shall hold more than one Class C and one Class D station license.

§ 95.14   Mailing address furnished by licensee.

Each application shall set forth and each licensee shall furnish the Commission with an address in the United States to be used by the Commission in serving documents or directing correspondence to that licensee. Unless any licensee advises the Commission to the contrary, the address contained in the licensee's most recent application will be used by the Commission for this purpose.

[*§ 95.14 added new eff. 2–5–75; VI (75)–1*]

§ 95.15   Filing of applications.

(a) To assure that necessary information is supplied in a consistent manner by all persons, standard forms are prescribed for use in connection with the majority of applications and reports submitted for Commission consideration. Standard numbered forms applicable to the Citizens Radio Service are discussed in § 95.19 and may be obtained from the Washington, D.C., 20554, office of the Commission, or from any of its engineering field offices.

(b) All formal applications for Class C or Class D new, modified, or renewal station authorizations shall be submitted to the Commission's Office at 334 York Street, Gettysburg, Pa. 17325. Applications for Class A station authorizations, applications for consent to transfer of control of a corporation holding any citizens radio station authorization, requests for special temporary authority or other special requests, and correspondence relating to an application for any class citizens radio station authorization shall be submitted to the Commission's Office at Washington, D.C. 20554, and should be directed to the attention of the Secretary. Beginning January 1, 1973, applicants for Class A stations in the Chicago Regional Area, defined in § 95.19, shall submit their applications to the Commission's Chicago Regional Office. The address of the Regional Office will be announced at a later date. Applications involving Class A or Class D station equipment which is neither type approved nor crystal controlled, whether of commercial or home construction, shall be accompanied by supplemental data describing in detail the design and construction of the transmitter and methods employed in testing it to determine compliance with the technical requirements set forth in Subpart C of this part.

(c) Unless otherwise specified, an application shall be filed at least 60 days prior to the date on which it is desired that Commission action thereon be completed. In any case where the applicant has made timely and sufficient application for renewal of license, in accordance with the Commission's rules, no license with reference to any activity of a continuing nature shall expire until such application shall have been finally determined.

(d) Failure on the part of the applicant to provide all the information required by the application form, or to supply the necessary exhibits or supplementary statements may constitute a defect in the application.

(e) Applicants proposing to construct a radio station on a site located on land under the jurisdiction of the U.S. Forest Service, U.S. Department of Agriculture, or the Bureau of Land Management, U.S. Department of the Interior, must supply the information and must follow the procedure prescribed by § 1.70 of this chapter.

§ 95.17   Who may sign applications.

(a) Except as provided in paragraph (b) of this section, applications, amendments thereto, and related statements of fact required by the Commission shall be personally signed by the applicant, if the applicant is an individual; by one of the partners, if the applicant is a partnership; by an officer, if the applicant is a corporation; or by a member who is an officer, if the applicant is an unincorporated association. Applications, amendments, and related statements of fact filed on behalf of eligible government entities, such as states and territories of the United States and political subdivisions thereof, the District of Columbia, and units of local government, including incorporated municipalities, shall be signed by such duly elected or appointed officials as may be competent to do so under the laws of the applicable jurisdiction.

(b) Applications, amendments thereto, and related statements of fact required by the Commission may be signed by the applicant's attorney in case of the applicant's physical disability or of his absence from the United States. The attorney shall in that event separately set forth the reason why the application is not signed by the applicant. In addition, if any matter is stated on the basis of the attorney's belief only (rather than his knowledge), he shall separately set forth his reasons for believing that such statements are true.

(c) Only the original of applications, amendments, or related statements of fact need be signed; copies may be conformed.

(d) Applications, amendments, and related statements of fact need not be signed under oath. Willful false statements made therein, however, are punishable by fine and imprisonment. U.S. Code, Title 18, section 1001, and by appropriate administrative sanctions, including revocation of station license pursuant to section 312(a) (1) of the Communications Act of 1934, as amended.

§ 95.19   Standard forms to be used.

(a) *FCC Form 505, Application for Class C or D Station License in the Citizens Radio Service.* This form shall be used when:

(1) Application is made for a new Class C or Class D authorization. A separate application shall be submitted for each proposed class of station.

(2) Application is made for modification of any existing Class C or Class D station authorization in those cases where prior Commission approval of certain changes is required (see § 95.35).

(3) Application is made for renewal of an existing Class C or Class D station authorization, or for reinstatement of such an expired authorization.

(b) *FCC Form 400, Application for Radio Station Authorization in the Safety and Special Radio Services.* Except as provided in paragraph (d) of this section, this form shall be used when:

(1) Application is made for a new Class A base station or fixed station authorization. Separate applications shall be submitted for each proposed base or fixed station at different fixed locations; however, all equipment intended to be operated at a single fixed location is considered to be one station which may, if necessary, be classed as both a base station and a fixed station.

(2) Application is made for a new Class A station authorization for any required number of mobile units (including hand-carried and pack-carried units) to be operated as a group in a single radiocommunication system in a particular area. An application for Class A mobile station authorization may be combined with the application for a single Class A base station authorization when such mobile units are to be operated with that base station only.

(3) Application is made for station license of any Class A base station or fixed station upon completion of construction or installation in accordance with the terms and conditions set

forth in any construction permit required to be issued for that station, or application for extension of time within which to construct such a station.

(4) Application is made for modification of any existing Class A station authorization in those cases where prior Commission approval of certain changes is required (see § 95.35).

(5) Application is made for renewal of an existing Class A station authorization, or for reinstatement of such an expired authorization.

(6) Each applicant in the Safety and Special Radio Services (1) for modification of a station license involving a site change or a substantial increase in tower height or (2) for a license for a new station must, before commencing construction, supply the environmental information, where required, and must follow the procedure prescribed by Subpart I of Part 1 of this chapter (§§ 1.1301 through 1.1319) unless Commission action authorizing such construction would be a minor action with the meaning of Subpart I of Part 1.

(7) Application is made for an authorization for a new Class A base or fixed station to be operated at unspecified or temporary loctions. When one or more individual transmitters are each intended to be operated as a base station or as a fixed station at unspecified or temporary locations for indeterminate periods, such transmitters may be considered to comprise a single station intended to be operated at temporary locations. The application shall specify the general geographic area within which the operation will be confined. Sufficient data must be submitted to show the need for the proposed area of operation.

(c) *FCC Form 703, Application for Consent to Transfer of Control of Corporation Holding Construction Permit or Station License.* This form shall be used when application is made for consent to transfer control of a corporation holding any citizens radio station authorization.

(d) Beginning April 1, 1972, FCC Form 425 shall be used in lieu of FCC Form 400, applicants for Class A stations located in the Chicago Regional Area defined to consist of the counties listed below:

ILLINOIS

1. Boone.	11. De Witt.
2. Bureau.	12. Douglas.
3. Carroll.	13. Du Page.
4. Champaign.	14. Edgar.
5. Christian.	15. Ford
6. Clark	16. Fulton.
7. Coles.	17. Grundy
8. Cook.	18. Henry.
9. Cumberland.	19. Iroquois.
10. De Kalb.	20. Jo Daviess.
21. Kane.	38. Ogle.
22. Kankakee.	39. Peoria.
23. Kendall.	40. Piatt.
24. Knox.	41. Putnam.
25. Lake.	42. Rock Island.
26. La Salle.	43. Sangamon.
27. Lee.	44. Shelby.
28. Livingston.	45. Stark.
29. Logan.	46. Stephenson.
30. Macon.	47. Tazewell.
31. Marshall.	48. Vermilion.
32. Mason.	49. Warren.
33. McHenry.	50. Whiteside.
34. McLean.	51. Will.
35. Menard.	52. Winnebago.
36. Mercer.	53. Woodford.
37. Moultrie.	

INDIANA

1. Adams.	28. Madison.
2. Allen.	29. Marion.
3. Benton.	30. Marshall.
4. Blackford.	31. Miami.
5. Boone.	32. Montgomer.
6. Carroll.	33. Morgan.
7. Cass.	34. Newton.
8. Clay.	35. Noble.
9. Clinton.	36. Owen.
10. De Kalb.	37. Parke.
11. Delaware.	38. Porter.
12. Elkhart.	39. Pulaski.
13. Fountain.	40. Putnam.
14. Fulton.	41. Randolph.
15. Grant.	42. St. Joseph.
16. Hamilton.	43. Starke.
17. Hancock.	44. Steuben.
18. Hendricks.	45. Tippecanoe.
19. Henry.	46. Tipton.
20. Howard.	47. Vermilion.
21. Huntington.	48. Vigo.
22. Jasper.	49. Wabash.
23. Jay.	50. Warren.
24. Kosciusko.	51. Wells.
25. Lake.	52. White.
26. Lagrange.	53. Whitley.
27. La Porte.	

IOWA

1. Cedar.	5. Jones.
2. Clinton.	6. Muscatine.
3. Dubuque.	7. Scott.
4. Jackson.	

MICHIGAN

1. Allegan.	6. Cass.
2. Barry.	7. Clinton.
3. Berrien.	8. Eaton.
4. Branch.	9. Hillsdale.
5. Calhoun.	10. Ingham.

11. Ionia.
12. Jackson.
13. Kalamazoo.
14. Kent.
15. Lake.
16. Mason.
17. Mecosta.

18. Montcalm.
19. Muskegon.
20. Newaygo.
21. Oceana.
22. Ottawa.
23. St. Joseph.
24. Van Buren.

OHIO

1. Defiance.
2. Mercer.
3. Paulding.

4. Van Wert.
5. Williams.

WISCONSIN

1. Adams.
2. Brown.
3. Calumet.
4. Columbia.
5. Dane.
6. Dodge.
7. Door.
8. Fond du Lac.
9. Grant.
10. Green.
11. Green Lake.
12. Iowa.
13. Jefferson.
14. Juneau.
15. Kenosha.
16. Kewaunee.
17. Lafayette.

18. Manitowoc.
19. Marquette.
20. Milwaukee.
21. Outagamie.
22. Ozaukee.
23. Racine.
24. Richland.
25. Rock.
26. Sauk.
27. Sheboygan.
28. Walworth.
29. Washington.
30. Waukesha.
31. Waupaca.
32. Waushara.
33. Winnebago.

§ 95.25   Amendment or dismissal of application.

(a) Any application may be amended upon request of the applicant as a matter of right prior to the time the application is granted or designated for hearing. Each amendment to an application shall be signed and submitted in the same manner and with the same number of copies as required for the original application.

(b) Any application may, upon written request signed by the applicant or his attorney, be dismissed without prejudice as a matter of right prior to the time the application is granted or designated for hearing.

§ 95.27   Transfer of license prohibited.

A station authorization in the Citizens Radio Service may not be transferred or assigned. In lieu of such transfer or assignment, an application for new station authorization shall be filed in each case, and the previous authorization shall be forwarded to the Commission for cancellation.

§ 95.29   Defective applications.

(a) If an applicant is requested by the Commission to file any documents or information not included in the prescribed application form,

a failure to comply with such request will constitute a defect in the application.

(b) When an application is considered to be incomplete or defective, such application will be returned to the applicant, unless the Commission may otherwise direct. The reason for return of the applications will be indicated, and if appropriate, necessary additions or corrections will be suggested.

§ 95.31   Partial grant.

Where the Commission, without a hearing, grants an application in part, or with any privileges, terms, or conditions other than those requested, the action of the Commission shall be considered as a grant of such application unless the applicant shall, within 30 days from the date on which such grant is made, or from its effective date if a later date is specified, file with the Commission a written rejection of the grant as made. Upon receipt of such rejection, the Commission will vacate its original action upon the application and, if appropriate, set the application for hearing.

§ 95.33   License term.

Licenses for stations in the Citizens Radio Service will normally be issued for a term of 5 years from the date of original issuance, major modification, or renewal.

§ 95.35   Changes in transmitters and authorized stations.

Authority for certain changes in transmitters and authorized stations must be obtained from the Commission before the changes are made, while other changes do not require prior Commission approval. The following paragraphs of this section describe the conditions under which prior Commission approval is or is not necessary.

(a) Proposed changes which will result in operation inconsistent with any of the terms of the current authorization require that an application for modification of license be submitted to the Commission. Application for modification shall be submitted in the same manner as an application for a new station license, and the licensee shall forward his existing authorization to the Commission for cancellation immediately upon receipt of the superseding authorization. Any of the following changes to authorized stations may be made only upon approval by the Commission:

(1) Increase the overall number of transmitters authorized.

(2) Change the presently authorized location of a Class A fixed or base station or control point.

(3) Move, change the height of, or erect a Class A station antenna structure.

(4) Make any change in the type of emission or any increase in bandwidth of emission or power of a Class A station.

(5) Addition or deletion of control point(s) for an authorized transmitter of a Class A station.

(6) Change or increase the area of operation of a Class A mobile station or a Class A base or fixed station authorized to be operated at temporary locations.

(7) Change the operating frequency of a Class A station.

(b) When the name of a licensee is changed (without changes in the ownership, control, or corporate structure), or when the mailing address of the licensee is changed (without changing the authorized location of the base or fixed Class A station) a formal application for modification of the license is not required. However, the licensee shall notify the Commission promptly of these changes. The notice, which may be in letter form, shall contain the name and address of the licensee as they appear in the Commission's records, the new name and/or address, as the case may be, and the call signs and classes of all radio stations authorized to the licensee under this part. The notice concerning Class C or D radio stations shall be sent to Federal Communications Commission, Gettysburg, Pa. 17325, and a copy shall be maintained with the records of the station. The notice concerning Class A stations shall be sent to (1) Secretary, Federal Communications Commission, Washington, D.C. 20554, and (2) to Engineer in Charge of the Radio District in which the station is located, and a copy shall be maintained with the license of the station until a new license is issued.

(c) Proposed changes which will not depart from any of the terms of the outstanding authorization for the station may be made without prior Commission approval. Included in such changes is the substitution of transmitting equipment at any station, provided that the equipment employed is included in the Commission's "Radio Equipment List," and is listed as acceptable for use in the appropriate class of station in this service. Provided it is crystal-controlled and otherwise complies with the power, frequency tolerance, emission and modulation percentage limitations prescribed, non-type accepted equipment may be substituted at:

(1) Class C stations operated on frequencies in the 26.99–27.26 MHz band;

(2) Class D stations until November 22, 1974.

(d) Transmitting equipment type accepted for use in Class D stations shall not be modified by the user. Changes which are specifically prohibited include:

(1) Internal or external connection or addition of any part, device or accessory not included by the manufacturer with the transmitter for its type acceptance. This shall not prohibit the external connection of antennas or antenna transmission lines, antenna switches, passive networks for coupling transmission lines or antennas to transmitters, or replacement of microphones.

(2) Modification in any way not specified by the transmitter manufacturer and not approved by the Commission.

(3) Replacement of any transmitter part by a part having different electrical characteristics and ratings from that replaced unless such part is specified as a replacement by the transmitter manufacturer.

(4) Substitution or addition of any transmitter oscillator crystal unless the crystal manufacturer or transmitter manufacturer has made an express determination that the crystal type, as installed in the specific transmitter type, will provide that transmitter type with the capability of operating within the frequency tolerance specified in Section 95.45 (a).

(5) Addition or substitution of any component, crystal or combination of crystals, or any other alteration to enable transmission on any frequency not authorized for use by the licensee.

(e) Only the manufacturer of the particular unit of equipment type accepted for use in Class D stations may make the permissive changes allowed under the provisions of Part 2 of this chapter for type acceptance. However, the manufacturer shall not make any of the following changes to the transmitter without prior written authorization from the Commission:

(1) Addition of any accessory or device not specified in the application for type acceptance and approved by the Commission in granting said type acceptance.

(2) Addition of any switch, control, or external connection.

(3) Modification to provide capability for an additional number of transmitting frequencies.

§ 95.37 Limitations on antenna structures.

(a) Except as provided in paragraph (b) of this section, an antenna for a Class A station which exceeds the following height limitations may not be erected or used unless notice has been filed with both the FAA on FAA Form 7460–1 and with the Commission on Form 714 or on the license application form, and prior approval by the Commission has been obtained for:

(1) Any construction or alteration of more

than 200 feet in height above ground level at its site (§ 17.7 (a) of this chapter).

(2) Any construction or alteration of greater height than an imaginary surface extending outward and upward at one of the following slopes (§ 17.7 (b) of this chapter):

(i) 100 to 1 for a horizontal distance of 20,000 feet from the nearest point of the nearest runway of each airport with at least one runway more than 3,200 feet in length, excluding heliports, and seaplane bases without specified boundaries, if that airport is either listed in the Airport Directory of the current Airman's Information Manual or is operated by a Federal military agency.

(ii) 50 to 1 for a horizontal distance of 10,000 feet from the nearest point of the nearest runway of each airport with its longest runway no more than 3,200 feet in length, excluding heliports, and seaplane bases without specified boundaries, if that airport is either listed in the Airport Directory or is operated by a Federal military agency.

(iii) 25 to 1 for a horizontal distance of 5,000 feet from the nearest point of the nearest landing and take-off area of each heliport listed in the Airport Directory or operated by a Federal military agency.

(3) Any construction or alteration on any airport listed in the Airport Directory of the current Airman's Information Manual (§ 17.7 (c) of this chapter).

(b) A notification to the Federal Aviation Administration is not required for any of the following construction or alteration of Class A station antenna structures.

(1) Any object that would be shielded by existing structures of a permanent and substantial character or by natural terrain or topographic features of equal or greater height, and would be located in the congested area of a city, town, or settlement where it is evident beyond all reasonable doubt that the structure so shielded will not adversely affect safety in air navigation. Applicants claiming such exemption shall submit a statement with their application to the Commission explaining the basis in detail for their finding (§ 17.14 (a) of this chapter).

(2) Any antenna structure of 20 feet or less in height except one that would increase the height of another antenna structure (§ 17.14 (b) of this chapter).

(c) A Class C or Class D station operated at a fixed location shall employ a transmitting antenna which complies with at least one of the following:

(1) The antenna and its supporting structure does not exceed 20 feet in height above ground level; or

(2) The antenna and its supporting structure does not exceed by more than 20 feet the height of any natural formation, tree or man-made structure on which it is mounted; or

NOTE: A man-made structure is any construction other than a tower, mast, or pole.

(3) The antenna is mounted on the transmitting antenna structure of another authorized radio station and does not exceed the height of the antenna supporting structure of the other station; or

(4) The antenna is mounted on and does not exceed the height of the antenna structure otherwise used solely for receiving purposes, which structure itself complies with subparagraph (1) or (2) of this paragraph.

(5) The antenna is omnidirectional and the highest point of the antenna and its supporting structure does not exceed 60 feet above ground level and the highest point also does not exceed one foot in height above the established airport elevation for each 100 feet of horizontal distance from the nearest point of the nearest airport runway.

NOTE: A work sheet will be made available upon request to assist in determining the maximum permissible height of an antenna structure.

(d) Class C stations operated on frequencies in the 72–76 MHz band shall employ a transmitting antenna which complies with all of the following:

(1) The gain of the antenna shall not exceed that of a half-wave dipole;

(2) The antenna shall be immediately attached to, and an integral part of, the transmitter; and

(3) Only vertical polarization shall be used.

(e) Further details as to whether an aeronautical study and/or obstruction marking and lighting may be required, and specifications for obstruction marking and lighting when required, may be obtained from Part 17 of this chapter, "Construction, Marking, and Lighting of Antenna Structures."

(f) Subpart I of Part 1 of this chapter contains procedures implementing the National Environmental Policy Act of 1969. Applications for authorization of the construction of certain classes of communications facilities defined as "major actions" in § 1.305 thereof, are required to be accompanied by specified statements. Generally these classes are:

(1) Antenna towers or supporting structures which exceed 300 feet in height and are not located in areas devoted to heavy industry or to agriculture.

(2) Communications facilities to be located in the following areas:

(i) Facilities which are to be located in an

officially designated wilderness area or in an area whose designation as a wilderness is pending consideration;

(ii) Facilities which are to be located in an officially designated wildlife preserve or in an area whose designation as a wildlife preserve is pending consideration;

(iii) Facilities which will affect districts, sites, buildings, structures or objects, significant in American history, architecture, archaeology or culture, which are listed in the National Register of Historic Places or are eligible for listing (see 36 CFR 800.2 (d) and (f) and 800.10); and

(iv) Facilities to be located in areas which are recognized either nationally or locally for their special scenic or recreational value.

(3) Facilities whose construction will involve extensive change in surface features (e.g. wetland fill, deforestation or water diversion).

NOTE: The provisions of this paragraph do not include the mounting of FM, television or other antennas comparable thereto in size on an existing building or antenna tower. The use of existing routes, buildings and towers is an environmentally desirable alternative to the construction of new routes or towers and is encouraged.

If the required statements do not accompany the application, the pertinent facts may be brought to the attention of the Commission by any interested person during the course of the license term and considered de novo by the Commission.

SUBPART C—
TECHNICAL REGULATIONS

§ 95.41 Frequencies available.

(a) Frequencies available for assignment to Class A stations:

(1) The following frequencies or frequency pairs are available primarily for assignment to base and mobile stations. They may also be assigned to fixed stations as follows:

(i) Fixed stations which are used to control base stations of a system may be assigned the frequency assigned to the mobile units associated with the base station. Such fixed stations shall comply with the following requirements if they are located within 75 miles of the center of urbanized areas of 200,000 or more population.

(a) If the station is used to control one or more base stations located within 45 degrees of azimuth, a directional antenna having a front-to-back ratio of at least 15 dB shall be used at the fixed station. For other situations where such a directional antenna cannot be used, a cardioid, bidirectional or omnidirectional antenna may be employed. Consistent with

reasonable design, the antenna used must, in each case, produce a radiation pattern that provides only the coverage necessary to permit satisfactory control of each base station and limit radiation in other directions to the extent feasible.

(b) The strength of the signal of a fixed station controlling a single base station may not exceed the signal strength produced at the antenna terminal of the base receiver by a unit of the associated mobile station, by more than 6 dB. When the station controls more than one base station, the 6 dB control-to-mobile signal difference need be verified at only one of the base station sites. The measurement of the signal strength of the mobile unit must be made when such unit is transmitting from the control station location or, if that is not practical, from a location within one-fourth mile of the control station site.

(c) Each application for a control station to be authorized under the provisions of this paragraph shall be accompanied by a statement certifying that the output power of the proposed station transmitter will be adjusted to comply with the foregoing signal level limitation. Records of the measurements used to determine the signal ratio shall be kept with the station records and shall be made available for inspection by Commission personnel upon request.

(d) Urbanized areas of 200,000 or more population are defined in the U.S. Census of Population, 1960, vol. 1, table 23, page 50. The centers of urbanized areas are determined from the Appendix, page 226 of the U.S. Commerce publication "Air Line Distance Between Cities in the United States."

(ii) Fixed stations, other than those used to control base stations, which are located 75 or more miles from the center of an urbanized area of 200,000 or more population. The centers of urbanized areas of 200,000 or more population are listed on page 226 of the Appendix to the U.S. Department of Commerce publication "Air Line Distance Between Cities in the United State." When the fixed station is located 100 miles or less from the center of such an urbanized area, the power output may not exceed 15 watts. All fixed systems are limited to a maximum of two frequencies and must employ directional antennas with a front-to-back ratio of at least 15 dB. For two-frequency systems, separation between transmit-receive frequencies is 5 MHz.

Base and Mobile (MHz)	Mobile Only (MHz)
462.550	467.550
462.575	467.575
462.600	467.600

Base and Mobile (MHz)	Mobile Only (MHz)
462.625	467.625
462.650	467.650
462.675	467.675
462.700	467.700
462.725	467.725

(2) Conditions governing the operation of stations authorized prior to March 18, 1968:

(i) All base and mobile stations authorized to operate on frequencies other than those listed in subparagraph (1) of this paragraph may continue to operate on those frequencies only until January 1, 1970.

(ii) Fixed stations located 100 or more miles from the center of any urbanized area of 200,000 or more population authorized to operate on frequencies other than those listed in subparagraph (1) of this paragraph will not have to change frequencies provided no interference is caused to the operation of stations in the land mobile service.

(iii) Fixed stations, other than those used to control base stations, located less than 100 miles (75 miles if the transmitter power output does not exceed 15 watts) from the center of any urbanized area of 200,000 or more population must discontinue operation by November 1, 1971. However, any operation after January 1, 1970, must be on frequencies listed in subparagraph (1) of this paragraph.

(iv) Fixed stations, located less than 100 miles from the center of any urbanized area of 200,000 or more population, which are used to control base stations and are authorized to operate on frequencies other than those listed in subparagraph (1) of this paragraph may continue to operate on those frequencies only until January 1, 1970.

(v) All fixed stations must comply with the applicable technical requirements of subparagraph (1) relating to antennas and radiated signal strength of this paragraph by November 1, 1971.

(vi) Notwithstanding the provisions of subdivisions (i) through (v) of this subparagraph, all stations authorized to operate on frequencies between 465.000 and 465.500 MHz and located within 75 miles of the center of the 20 largest urbanized areas of the United States, may continue to operate on these frequencies only until January 1, 1969. An extension to continue operation on such frequencies until January 1, 1970, may be granted to such station licensees on a case by case basis if the Commission finds that continued operation would not be inconsistent with planned usage of the particular frequency for police purposes. The 20 largest urbanized areas can be found in the U.S. Census

of Population, 1960, vol. 1, table 23, page 50. The centers of urbanized areas are determined from the appendix, page 226, of the U.S. Commerce publication, "Air Line Distance Between Cities in the United States."

(b) [Reserved]

(c) Class C mobile stations may employ only amplitude tone modulation or on-off keying of the unmodulated carrier, on a shared basis with other stations in the Citizens Radio Service on the frequencies and under the conditions specified in the following tables:

(1) For the control of remote objects or devices by radio, or for the remote actuation of devices which are used solely as a means of attracting attention and subject to no protection from interference due to the operation of industrial, scientific, or medical devices within the 26.96–27.28 MHz band, the following frequencies are available:

(MHz)	(MHz)	(MHz)
26.995	27.095	27.195
27.045	27.145	[1]27.255

[1]The frequency 27.255 MHz also is shared with stations in other services.

(2) Subject to the conditions that interference will not be caused to the remote control of industrial equipment operating on the same or adjacent frequencies and to the reception of television transmissions on Channels 4 or 5; and that no protection will be afforded from interference due to the operation of fixed and mobile stations in other services assigned to the same or adjacent frequencies in the band, the following frequencies are available solely for the radio remote control of models used for hobby purposes:

(i) For the radio remote control of any model used for hobby purposes:

MHz	MHz	MHz
72.16	72.32	72.96

(ii) For the radio remote control of aircraft models only:

MHz	MHz	MHz
72.08	72.24	72.40
75.64		

(d) The frequencies listed in the following tables are available for use by Class D mobile stations employing radiotelephony only, on a shared basis with other stations in the Citizens Radio Service, and subject to no protection from interference due to the operation of industrial, scientific, or medical devices within the 26.96–27.28 MHz band.

(1) The following frequencies, commonly known as Channels 1 through 8 and 10 through 23, may be used for communications between units of the same station:

MHz	CHANNELS
26.965	1
26.975	2
26.985	3
27.005	4
27.015	5
27.025	6
27.035	7
27.055	8
27.075	10
27.085	11
27.105	12
27.115	13
27.125	14
27.135	15
27.155	16
27.165	17
27.175	18
27.185	19
27.205	20
27.215	21
27.225	22
27.255	23

(2) Only the following frequencies may be used for communications between units of different stations:

MHz	Channel
27.075	10
27.085	11
27.105	12
27.115	13
27.125	14
27.135	15
27.255	23

(3) The frequency 27.065 MHz (Channel 9) shall be used solely for:

(i) Emergency communications involving the immediate safety of life of individuals or the immediate protection of property or

(ii) Communications necessary to render assistance to a motorist.

NOTE: A licensee, before using Channel 9, must make a determination that his communication is either or both (a) an emergency communication or (b) is necessary to render assistance to a motorist. To be an emergency communication, the message must have some direct relation to the immediate safety of life or immediate protection of property. If no immediate action is required, it is not an emergency. What may not be an emergency under one set of circumstances may be an emergency under different circumstances. There are many worthwhile public service communications that do not qualify as emergency communications. In the case of motorist assistance, the message must be necessary to assist a particular motorist and not, except in a valid emergency, motorists in general. If the communications are to be lengthy, the exchange should be shifted to another channel, if feasible, after contact is established. No nonemergency or nonmotorist assistance communications are permitted on Channel 9 even for the limited purpose of calling a licensee monitoring a channel to ask him to switch to another channel. Although Channel 9 may be used for marine emergencies, it should not be considered a substitute for the authorized marine distress system. The Coast Guard has stated it will not "participate directly in the Citizens Radio Service by fitting with and/or providing a watch on any Citizens Band Channel. (Coast Guard Commandant Instruction 2302.6.)"

The following are examples of permitted and prohibited types of communications. They are guidelines and are not intended to be all inclusive.

Permitted   *Example message*

Yes——"A tornado sighted six miles north of town."

No——"This is observation post number 10. No tornados sighted."

Yes——"I am out of gas on Interstate 95."

No——"I am out of gas in my driveway."

Yes——"There is a four-car collision at Exit 10 on the Beltway, send police and ambulance."

No——"Traffic is moving smoothly on the Beltway."

Yes——"Base to Unit 1, the Weather Bureau has just issued a thunderstorm warning. Bring the sailboat into port."

No——"Attention all motorists. The Weather Bureau advises that the snow tomorrow will accumulate 4 to 6 inches."

Yes——"There is a fire in the building on the corner of 6th and Main Streets."

No——"This is Halloween patrol unit number 3. Everything is quiet here."

The following priorities should be observed in the use of Channel 9.

1. Communications relating to an existing situation dangerous to life or property, i.e., fire, automobile accident.

2. Communications relating to a potentially hazardous situation, i.e., car stalled in a dangerous place, lost child, boat out of gas.

3. Road assistance to a disabled vehicle on the highway or street.

4. Road and street directions.

(e) Upon specific request accompanying application for renewal of station authorization, a Class A station in this service, which was authorized to operate on a frequency in the 460--461 MHz band until March 31, 1967, may be assigned that frequency for continued use until not later than March 31, 1968, subject to all other provisions of this part.

§ 95.43 Transmitter power.

(a) Transmitter power is the power at the transmitter output terminals and delivered to the antenna, antenna transmission line, or any other impedance-matched, radio frequency load.

(1) For single sideband transmitters and other transmitters employing a reduced carrier, a suppressed carrier or a controlled carrier, used at Class D stations, transmitter power is the peak envelope power.

(2) For all transmitters other than those covered by paragraph (a)(1) of this section, the transmitter power is the carrier power.

(b) The transmitter power of a station shall not exceed the following values under any condition of modulation or other circumstances.

Class of station:	Transmitter power in watts
A	50
C—27.255 MHz	25
C—26.995–27.195 MHz	4
C—72–76 MHz	0.75
D—Carrier (where applicable)	4
D—Peak envelope power (where applicable)	12

§ 95.44 External radio frequency power amplifiers prohibited.

No external radio frequency power amplifier shall be used or attached, by connection, coupling attachment or in any other way at any Class D station.

NOTE: An external radio frequency power amplifier at a Class D station will be presumed to have been used where it is in the operator's possession or on his premises and there is extrinsic evidence of any operation of such Class D station in excess of power limitations provided under this rule part unless the operator of such equipment holds a station license in another radio service under which license the use of the said amplifier at its maximum rated output power is permitted.

§ 95.45 Frequency tolerance.

(a) Except as provided in paragraphs (b) and (c) of this section, the carrier frequency of a transmitter in this service shall be maintained within the following percentage of the authorized frequency:

Class of station	Frequency tolerance	
	Fixed and base	Mobile
A	0.00025	0.0005
C		.005
D		.005

(b) Transmitters used at Class C stations operating on authorized frequencies between 26.99 and 27.26 MHz with 2.5 watts or less mean output power, which are used solely for the control of remote objects or devices by radio (other than devices used solely as a means of attracting attention), are permitted a frequency tolerance of 0.01 percent.

(c) Class A stations operated at a fixed location used to control base stations, through use of a mobile only frequency, may operate with a frequency tolerance of 0.0005 percent.

§ 95.47 Types of emission.

(a) Except as provided in paragraph (e) of this section, Class A stations in this service will normally be authorized to transmit radiotelephony only. However, the use of tone signals or signaling devices solely to actuate receiver circuits, such as tone operated squelch or selective calling circuits, the primary function of which is to establish or establish and maintain voice communications, is permitted. The use of tone signals solely to attract attention is prohibited.

(b) [Reserved]

(c) Class C stations in this service are authorized to use amplitude tone modulation or on-off unmodulated carrier only, for the control of remote objects or devices by radio or for the remote actuation of devices which are used solely as a means of attracting attention. The transmission of any form of telegraphy, telephony or record communications by a Class C station is prohibited. Telemetering, except for the transmission of simple, short duration signals indicating the presence or absence of a condition or the occurrence of an event, is also prohibited.

(d) Transmitters used at Class D stations in this service are authorized to use amplitude voice modulation, either single or double sideband. Tone signals or signalling devices may be used only to actuate receiver circuits, such as tone operated squelch or selective calling circuits, the primary function of which is to establish or maintain voice communications.

The use of any signals solely to attract attention or for the control of remote objects or devices is prohibited.

(e) Other types of emission not described in paragraph (a) of this section may be authorized for Class A citizens radio stations upon a showing of need therefor. An application requesting such authorization shall fully describe the emission desired, shall indicate the bandwidth required for satisfactory communication, and shall state the purpose for which such emission is required. For information regarding the classification of emissions and the calculation of bandwidth, reference should be made to Part 2 of this chapter.

§ 95.49 Emission limitations.

(a) Each authorization issued to a Class A citizens radio station will show, as a prefix to the classification of the authorized emission, a figure specifying the maximum bandwidth to be occupied by the emission.

(b) [Reserved]

(c) The authorized bandwidth of the emission of any transmitter employing amplitude modulation shall be 8 kHz for double sideband, 4 kHz for single sideband and the authorized bandwidth of the emission of transmitters employing frequency or phase modulation (Class F2 for F3) shall be 20 kHz. The use of Class F2 and F3 emissions in the frequency band 26.96–27.28 MHz is not authorized.

(d) The mean power of emissions shall be attenuated below the mean power of the transmitter in accordance with the following schedule:

(1) When using emissions other than single sideband:

(i) On any frequency removed from the center of the authorized bandwidth by more than 50 percent up to and including 100 percent of the authorized bandwidth: At least 25 decibels;

(ii) On any frequency removed from the center of the authorized bandwidth by more than 100 percent up to and including 250 percent of the authorized bandwidth: At least 35 decibels;

(2) When using single sideband emissions:

(i) On any frequency removed from the center of the authorized bandwidth by more than 50 percent up to and including 150 percent of the authorized bandwidth: At least 25 decibels;

(ii) On any frequency removed from the center of the authorized bandwidth by more than 150 percent up to and including 250 percent of the authorized bandwidth: At least 35 decibels;

(3) On any frequency removed from the center of the authorized bandwidth by more than 250 percent of the authorized bandwidth: At least 43 plus 10 log10 (mean power in watts) decibels.

(e) When an unauthorized emission results in harmful interference, the Commission may, in its descretion, require appropriate technical changes in equipment to alleviate the interference.

§ 95.51 Modulation requirements.

(a) When double sideband, amplitude modulation is used for telephony, the modulation percentage shall be sufficient to provide efficient communication and shall not exceed 100 percent.

(b) Each transmitter for use in Class D stations, other than single sideband, suppressed carrier, or controlled carrier, for which type acceptance is requested after May 24, 1974, having more than 2.5 watts maximum output power shall be equipped with a device which automatically prevents modulation in excess of 100 percent on positive and negative peaks.

(c) The maximum audio frequency required for satisfactory radiotelephone intelligibility for use in this service is considered to be 3000 Hz.

(d) Transmitters for use at Class A stations shall be provided with a device which automatically will prevent greater than normal audio level from causing modulation in excess of that specified in this subpart; *Provided, however,* That the requirements of this paragraph shall not apply to transmitters authorized at mobile stations and having an output power of 2.5 watts or less.

(e) Each transmitter of a Class A station which is equipped with a modulation limiter in accordance with the provisions of paragraph (d) of this section shall also be equipped with an audio low-pass filter. This audio low-pass filter shall be installed between the modulation limiter and the modulated stage and, at audio frequencies between 3 kHz and 20 kHz, shall have an attenuation greater than the attenuation at 1 kHz by at least:

$$60 \log 10 \ (f/3) \ \text{decibels}$$

where "f" is the audio frequency in kHz. At audio frequencies above 20 kHz, the attenuation shall be at least 50 decibels greater than the attenuation at 1 kHz.

(f) Simultaneous amplitude modulation and frequency or phase modulation of a transmitter is not authorized.

(g) The maximum frequency deviation of frequency modulated transmitters used at Class A stations shall not exceed ±5 kHz.

§ 95.53 Compliance with technical requirements.

(a) Upon receipt of notification from the Commission of a deviation from the technical requirements of the rules in this part, the radiations of the transmitter involved shall be suspended immediately, except for necessary tests and adjustments, and shall not be resumed until such deviation has been corrected.

(b) When any citizens radio station licensee receives a notice of violation indicating that the station has been operated contrary to any of the provisions contained in Subpart C of this part, or where it otherwise appears that operation of a station in this service may not be in accordance with applicable technical standards, the Commission may require the licensee to conduct such tests as may be necessary to determine whether the equipment is capable of meeting these standards and to make such adjustments as may be necessary to assure compliance therewith. A licensee who is notified that he is required to conduct such tests and/or make adjustments must, within the time limit specified in the notice, report to the Commission the results thereof.

(c) All tests and adjustments which may be required in accordance with paragraph (b) of this section shall be made by, or under the immediate supervision of, a person holding a first- or second-class commercial operator license, either radiotelephone or radio telegraph as may be appropriate for the type of emission employed. In each case, the report which is submitted to the Commission shall be signed by the licensed commercial operator. Such report shall describe the results of the tests and adjustments, the test equipment and procedures used, and shall state the type, class, and serial number of the operator's license. A copy of this report shall also be kept with the station records.

§ 95.55 Acceptability of transmitters for licensing.

Transmitters type approved or type accepted for use under this part are included in the Commission's Radio Equipment List. Copies of this list are available for public reference at the Commission's Washington, D.C., offices and field offices. The requirements for transmitters which may be operated under a license in this service are set forth in the following paragraphs.

(a) Class A stations: All transmitters shall be type accepted.

(b) Class C stations:

(1) Transmitters operated in the band 72–76 MHz shall be typed accepted.

(2) All transmitters operated in the band 26.99–27.26 MHz shall be type approved, type accepted or crystal controlled.

(c) Class D stations:

(1) All transmitters first licensed, or marketed as specified in § 2.805 of this chapter, prior to November 22, 1974, shall be type accepted or crystal controlled.

(2) All transmitters first licensed, or marketed as specified in § 2.803 of this chapter, on or after November 22, 1974, shall be type accepted.

(3) Effective November 23, 1978, all transmitters shall be type accepted.

(4) Transmitters which are equipped to operate on any frequency not included in § 95.41 (d)(1) may not be installed at, or used by, any Class D station unless there is a station license posted at the transmitter location, or a transmitter identification card (FCC Form 452–C) attached to the transmitter, which indicates that operation of the transmitter on such frequency has been authorized by the Commission.

(d) With the exception of equipment type approved for use at a Class C station, all transmitting equipment authorized in this service shall be crystal controlled.

(e) No controls, switches or other functions which can cause operation in violation of the technical regulations of this part shall be accessible from the operating panel or exterior to the cabinet enclosing a transmitter authorized in this service.

§ 95.57 Procedure for type acceptance of equipment.

(a) Any manufacturer of a transmitter built for use in this service, except noncrystal controlled transmitters for use at Class C stations, may request type acceptance for such transmitter in accordance with the type acceptance requirements of this part, following the type acceptance procedure set forth in Part 2 of this chapter.

(b) Type acceptance for an individual transmitter may also be requested by an applicant for a station authorization by following the type acceptance procedures set forth in Part 2 of this chapter. Such transmitters, if accepted, will not normally be included on the Commission's "Radio Equipment List", but will be individually enumerated on the station authorization.

(c) Additional rules with respect to type acceptance are set forth in Part 2 of this chapter. These rules include information with respect to withdrawal of type acceptance, modification of type-accepted equipment, and limitations on the findings upon which type acceptance is based.

(d) Transmitters equipped with a frequency or frequencies not listed in § 95.41 (d)(1) will not be type accepted for use at Class D stations unless the transmitter is also type accepted for use in the service in which the frequency is authorized, if type acceptance in that service is required.

§ 95.58 Additional requirements for type acceptance.

(a) All transmitters shall be crystal controlled.

(b) Except for transmitters type accepted for use at Class A stations, transmitters shall not include any provisions for increasing power to levels in excess of the pertinent limits specified in Section 95.43.

(c) In addition to all other applicable technical requirements set forth in this part, transmitters for which type acceptance is requested after May 24, 1974, for use at Class D stations shall comply with the following:

(1) Single sideband transmitters and other transmitters employing reduced, suppressed or controlled carrier shall include a means for automatically preventing the transmitter power from exceeding either the maximum permissible peak envelope power or the rated peak envelope power of the transmitter, whichever is lower.

(2) Multi-frequency transmitters shall not provide more than 23 transmitting frequencies, and the frequency selector shall be limited to a single control.

(3) Other than the channel selector switch, all transmitting frequency determining circuitry, including crystals, employed in Class D station equipment shall be internal to the equipment and shall not be accessible from the exterior of the equipment cabinet or operating panel.

(4) Single sideband transmitters shall be capable of transmitting on the upper sideband. Capability for transmission also on the lower sideband is permissible.

(5) The total dissipation ratings, established by the manufacturer of the electron tubes or semiconductors which supply radio frequency power to the antenna terminals of the transmitter, shall not exceed 10 watts. For electron tubes, the rating shall be the Intermittent Commercial and Amateur Service (ICAS plate dissipation value if established. For semiconductors, the rating shall be the collector or device dissipation value, whichever is greater, which may be temperature de-rated to not more than 50°C.

(d) Only the following external transmitter controls, connections or devices will normally be permitted in transmitters for which type acceptance is requested after May 24, 1974, for use at Class D stations. Approval of additional controls, connections or devices may be given after consideration of the function to be performed by such additions.

(1) Primary power connection. (Circuitry or devices such as rectifiers, transformers, or inverters which provide the nominal rated transmitter primary supply voltage may be used without voiding the transmitter type acceptance.)

(2) Microphone connection.

(3) Radio frequency output power connection.

(4) Audio frequency power amplifier output connector and selector switch.

(5) On-off switch for primary power to transmitter. May be combined with receiver controls such as the receiver on-off switch and volume control.

(6) Upper-lower sideband selector; for single sideband transmitters only.

(7) Selector for choice of carrier level; for single sideband transmitters only. May be combined with sideband selector.

(8) Transmitting frequency selector switch.

(9) Transmit-receive switch.

(10) Meter(s) and selector switch for monitoring transmitter performance.

(11) Pilot lamp or meter to indicate the presence of radio frequency output power or that transmitter control circuits are activated to transmit.

(e) An instruction book for the user shall be furnished with each transmitter sold and one copy (a draft or preliminary copy is acceptable providing a final copy is furnished when completed) shall be forwarded to the Commission with each request for type acceptance or type approval. The book shall contain all information necessary for the proper installation and operation of the transmitter including:

(1) Instructions concerning all controls, adjustments and switches which may be operated or adjusted without causing violation of technical regulations of this part;

(2) Warnings concerning any adjustment which, according to the rules of this part, may be made only by, or under the immediate supervision of, a person holding a commercial first or second class radio operator license;

(3) Warnings concerning the replacement or substitution of crystals, tubes or other components which could cause violation of the technical regulations of this part and of the type acceptance or type approval requirements of Part 2 of this chapter.

(4) Warnings concerning licensing requirements and details concerning the application procedures for licensing.

§ 95.59 Submission of noncrystal controlled Class C station transmitters for type approval.

Type approval of noncrystal controlled transmitters for use at Class C stations in this service may be requested in accordance with the procedure specified in Part 2 of this chapter.

§ 95.61   Type approval of receiver-transmitter combinations.

Type approval will not be issued for transmitting equipment for operation under this part when such equipment is enclosed in the same cabinet, is constructed on the same chassis in whole or in part, or is identified with a common type or model number with a radio receiver, unless such receiver has been certificated to the Commission as complying with the requirements of Part 15 of this chapter.

§ 95.63   Minimum equipment specifications.

Transmitters submitted for type approval in this service shall be capable of meeting the technical specifications contained in this part, and in addition, shall comply with the following:

(a) Any basic instructions concerning the proper adjustment, use, or operation of the equipment that may be necessary shall be attached to the equipment in a suitable manner and in such positions as to be easily read by the operator.

(b) A durable nameplate shall be mounted on each transmitter showing the name of the manufacturer, the type or model designation, and providing suitable space for permanently displaying the transmitter serial number, FCC type approval number, and the class of station for which approved.

(c) The transmitter shall be designed, constructed, and adjusted by the manufacturer to operate on a frequency or frequencies available to the class of station for which type approval is sought. In designing the equipment, every reasonable precaution shall be taken to protect the user from high voltage shock and radio frequency burns. Connections to batteries (if used) shall be made in such a manner as to permit replacement by the user without causing improper operation of the transmitter. Generally accepted modern engineering principles shall be utilized in the generation of radio frequency currents so as to guard against unnecessary interference to other services. In cases of harmful interference arising from the design, construction, or operation of the equipment, the Commission may require appropriate technical changes in equipment to alleviate interference.

(d) Controls which may effect changes in the carrier frequency of the transmitter shall not be accessible from the exterior of any unit unless such accessibility is specifically approved by the Commission.

§ 95.65   Test procedure.

Type approval tests to determine whether radio equipment meets the technical specifications contained in this part will be conducted under the following conditions:

(a) Gradual ambient temperature variations from 0° to 125° F.

(b) Relative ambient humidity from 20 to 95 percent. This test will normally consist of subjecting the equipment for at least three consecutive periods of 24 hours each, to a relative ambient humidity of 20, 60, and 95 percent, respectively, at a temperature of approximately 80° F.

(c) Movement of transmitter or objects in the immediate vicinity thereof.

(d) Power supply voltage variations normally to be encountered under actual operating conditions.

(e) Additional tests as may be prescribed, if considered necessary or desirable.

§ 95.67   Certificate of type approval.

A certificate or notice of type approval, when issued to the manufacturer of equipment intended to be used or operated in the Citizens Radio Service, constitutes a recognition that on the basis of the test made, the particular type of equipment appears to have the capability of functioning in accordance with the technical specifications and regulations contained in this part: *Provided,* That all such additional equipment of the same type is properly constructed, maintained, and operated: *And provided further,* That no change whatsoever is made in the design or construction of such equipment except upon specific approval by the Commission.

SUBPART D—STATION OPERATING REQUIREMENTS

§ 95.83   Prohibited uses.

(a) A Citizens radio station shall not be used:

(1) For engaging in radio communications as a hobby or diversion, i.e., operating the radio station as an activity in and of itself.

NOTE: The following are typical, but not all inclusive, examples of the types of communications evidencing a use of Citizens radio as a hobby or diversion which are prohibited under this rule:

"You want to give me your handle and I'll ship you out a card the first thing in the morning;" or "Give my your 10–20 so I can ship you some wallpaper." (Communications to other licensees for the purpose of exchanging so-called "QSL" cards.)

"I'm just checking to see who is on the air."

"Just calling to see if you can hear me. I'm at Main and Broadway."

"Just heard your call sign and thought I'd like to get acquainted;" or "Just passing through and heard your call sign so I thought I'd give you a shout."

"Just sitting here copying the mail and thought I'd give you a call to see how you were doing." (Referring to an intent to communicate based solely on hearing another person engaged in the use of his radio.)

"My 10–20 is Main and Broad Streets. Thought I'd call so I can see how well this new rig is getting out."

"Got a new mike on this rig and thought I'd give you a call to find out how my modulation is."

"Just thought I would give you a shout and let you known I am still around. Thanks for coming back."

"Clear with Venezuela. Just thought I'd let you know I was copying you up here."

"Thought I'd give you a shout and see if you knew where the unmodulated carrier was coming from."

"Just thought I'd give you a call to find out how the skip is coming in over at your location."

"Go ahead breaker. What kind of a rig are you using? Come back with your 10–20."

(2) For any purpose, or in connection with any activity, which is contrary to Federal, State, or local law.

(3) For the transmission of communications containing obscene, indecent, or profane words, language, or meaning.

(4) To carry communications for hire, whether the remuneration or benefit received is direct or indirect.

(5) To communicate with stations authorized or operated under the provisions of other parts of this chapter, with unlicensed stations, or with U.S. Government or foreign stations (other than as provided in Subpart E of this part) except for communications pursuant to §§ 95.85(b) and 95.121 and, in the case of Class A stations, for communications with U.S. Government stations in those cases which require cooperation or coordination of activities.

(6) For any communication not directed to specific stations or persons, except for: (i) Emergency and civil defense communications as provided in §§ 95.85(b) and 95.121, respectively, (ii) test transmissions pursuant to § 95.93, and (iii) communications from a mobile unit to other units or stations for the sole purpose of requesting routing directions, assistance to disabled vehicles or vessels, informa-

tion concerning the availability of food or lodging, or any other assistance necessary to a licensee in transit.

(7) To convey program material for retransmission, live or delayed, on a broadcast facility.

NOTE: A Class A or Class D station may be used in connection with the administrative, engineering, or maintenance activities of a broadcasting station; a Class A or Class C station may be used for control functions by radio which do not involve the transmission of program material; and a Class A or Class D station may be used in the gathering of news items or preparation of programs: *Provided,* That the actual or recorded transmissions of the Citizens radio station are not broadcast at any time in whole or in part.

(8) To interfere maliciously with the communications of another station.

(9) For the direct transmission of any material to the public through public address systems or similar means.

(10) To transmit superfluous communications, i.e., any transmissions which are not necessary to communications which are permissible.

(11) For the transmission of music, whistling, sound effects, or any material for amusement or entertainment purposes, or solely to attract attention.

(12) To transmit the word "MAYDAY" or other international distress signals, except when a ship, aircraft, or other vehicle is threatened by grave and imminent danger and requests immediate assistance.

(13) For transmitting communications to stations of other licensees which relate to the technical performance, capabilities, or testing of any transmitter or other radio equipment, including transmissions concerning the signal strength or frequency stability of a transmitter, except as necessary to establish or maintain the specific communication.

(14) For relaying messages or transmitting communications for a person other than the licensee or members of his immediate family except: (i) Communications transmitted pursuant to §§ 95.85(b), 95.87(b)(7), and 95.121; (ii) upon specific prior Commission approval, communications between Citizens radio service stations at fixed locations where public telephone service is not provided; and (iii) communications reporting locally observed traffic conditions directed to persons engaged directly or indirectly in furnishing traffic condition information to the motoring public via broadcast facilities.

(15) For advertising or soliciting the sale of any goods or services.

(16) For transmitting messages in other than plain language. Abbreviations, including nationally or internationally recognized operating signals, may be used only if a list of all such abbreviations and their meaning is kept in the station records and made available to any Commission representative on demand.

(b) A Class D station may not be used to communicate with, or attempt to communicate with, any unit of the same or another station over a distance of more than 150 miles.

(c) A licensee of a Citizens radio station who is engaged in the business of selling Citizens radio transmitting equipment shall not allow a customer to operate under his station license. In addition, all communications by the licensee for the purpose of demonstrating such equipment shall consist only of brief messages addressed to other units of the same station.

§ 95.85 Emergency and assistance to motorist use.

(a) All Citizens radio stations shall give priority to the emergency communications of other stations which involve the immediate safety of life of individuals or the immediate protection of property.

(b) Any station in this service may be utilized during an emergency involving the immediate safety of life of individuals or the immediate protection of property for the transmission of emergency communications. It may also be used to transmit communications necessary to render assistance to a motorist.

(1) When used for transmission of emergency communications certain provisions in this part concerning use of frequencies (§ 95.41(d)); prohibited uses (§ 95.83(a) (5), (6), and (14)); operation by or on behalf of persons other than the licensee (§ 95.87); and duration of transmissions (§ 95.91(a) and (b)) shall not apply.

(2) When used for transmission of communications necessary to render assistance to a motorist, the provisions of this part concerning directing communications to specific persons or stations (§ 95.83(a) (6)); transmitting messages for other persons (§ 95.83(a) (14)); and duration of transmissions (§ 95.91 (b)) shall not apply.

(3) The exemptions granted from certain rule provisions in subparagraphs (1) and (2) of this paragraph may be rescinded by the Commission at its discretion.

(c) If the emergency use under paragraph (b) of this section extends over a period of 12 hours or more, notice shall be sent to the Commission in Washington, D.C., as soon as it is evident that the emergency has or will exceed 12 hours. The notice should include the identity of the stations participating, the nature of the emergency, and the use made of the stations. A single notice covering all participating stations may be submitted.

§ 95.87 Operation by, or on behalf of, persons other than the licensee.

(a) Transmitters authorized in this service must be under the control of the licensee at all times. A licensee shall not transfer, assign, or dispose of, in any manner, directly or indirectly, the operating authority under his station license, and shall be responsible for the proper operation of all units of the station.

(b) Citizens radio stations may be operated only by the following persons, except as provided in paragraph (c) of this section:

(1) The licensee;

(2) Members of the licensee's immediate family living in the same household;

(3) The partners, if the licensee is a partnership, provided the communications relate to the business of the partnership;

(4) The members, if the licensee is an unincorporated association, provided the communications relate to the business of the association;

(5) Employees of the licensee only while acting within the scope of their employment;

(6) Any person under the control or supervision of the licensee when the station is used solely for the control of remote objects or devices, other than devices used only as a means of attracting attention; and

(7) Other persons, upon specific prior approval of the Commission shown on or attached to the station license, under the following circumstances:

(i) Licensee is a corporation and proposes to provide private radiocommunication facilities for the transmission of messages or signals by or on behalf of its parent corporation, another subsidiary of the parent corporation, or its own subsidary. Any remuneration or compensation received by the licensee for the use of the radiocommunication facilities shall be governed by a contract entered into by the parties concerned and the total of the compensation shall not exceed the cost of providing the facilities. Records which show the cost of service and its nonprofit or cost-sharing basis shall be maintained by the licensee.

(ii) Licensee proposes the shared or cooperative use of a Class A station with one or more other licensees in this service for the purpose of communicating on a regular basis with units of their respective Class A stations, or with units of other Class A stations if the communications transmitted are otherwise permissible. The use of these private radiocommunication facilities shall be conducted pursuant to a written contract which shall provide that contributions to capital and operating expense shall be made on a nonprofit, cost-sharing basis, the cost to be divided on an equitable basis among all parties to the agreement. Records which show the cost of service and its nonprofit, cost-sharing basis shall be maintained by the licensee. In any case, however, licensee must show a separate and independent need for the particular units proposed to be shared to fulfill his own communications requirements.

(iii) Other cases where there is a need for other persons to operate a unit of licensee's radio station. Requests for authority may be made either at the time of the filing of the application for station license or thereafter by letter. In either case, the licensee must show the nature of the proposed use and that it relates to an activity of the licensee, how he proposes to maintain control over the transmitters at all times, and why it is not appropriate for such other person to obtain a station license in his own name. The authority, if granted, may be specific with respect to the names of the persons who are permitted to operate, or may authorize operation by unnamed persons for specific purposes. This authority may be revoked by the Commission, in its discretion, at any time.

(c) An individual who was formerly a citizens radio station licensee shall not be permitted to operate any citizens radio station of the same class licensed to another person until such time as he again has been issued a valid radio station license of that class, when his license has been:

(1) Revoked by the Commission.

(2) Surrendered for cancellation after the institution of revocation proceedings by the Commission.

(3) Surrendered for cancellation after a notice of apparent liability to forfeiture has been served by the Commission.

§ 95.89 Telephone answering services.

(a) Notwithstanding the provisions of § 95.87, a licensee may install a transmitting unit of his station on the premises of a telephone answering service. The same unit may not be operated under the authorization of more than one licensee. In all cases, the licensee must enter into a written agreement with the answering service. This agreement must be kept with the licensee's station records and must provide, as a minimum, that:

(1) The licensee will have control over the operation of the radio unit at all times;

(2) The licensee will have full and unrestricted access to the transmitter to enable him to carry out his responsibilities under his license;

(3) Both parties understand that the licensee

is fully responsible for the proper operation of the citizens radio station; and

(4) The unit so furnished shall be used only for the transmission of communications to other units belonging to the licensee's station.

(b) A citizens radio station licensed to a telephone answering service shall not be used to relay messages or transmit signals to its customers.

§ 95.91 Duration of transmissions.

(a) All communications or signals, regardless of their nature, shall be restricted to the minimum practicable transmission time. The radiation of energy shall be limited to transmissions modulated or keyed for actual permissible communications, tests, or control signals. Continuous or uninterrupted transmissions from a single station or between a number of communicating stations is prohibited, except for communications involving the immediate safety of life or property.

(b) Communications between or among Class D stations shall not exceed 5 consecutive minutes. At the conclusion of this 5-minute period, or upon termination of the exchange if less than 5 minutes, the station transmitting and the stations participating in the exchange shall remain silent for a period of at least 5 minutes and monitor the frequency or frequencies involved before any further transmissions are made. However, for the limited purpose of acknowledging receipt of a call, such a station or stations may answer a calling station and request that it stand by for the duration of the silent period. The time limitations contained in this paragraph may not be avoided by changing the operating frequency of the station and shall apply to all the transmissions of an operator who, under other provisions of this part, may operate a unit of more than one citizens radio station.

(c) The transmission of audible tone signals or a sequence of tone signals for the operation of the tone operated squelch or selective calling circuits in accordance with § 95.47 shall not exceed a total of 15 seconds duration. Continuous transmission of a subaudible tone for this purpose is permitted. For the purposes of this section, any tone or combination of tones having no frequency above 150 hertz shall be considered subaudible.

(d) The transmission of permissible control signals shall be limited to the minimum practicable time necessary to accomplish the desired control or actuation of remote objects or devices. The continuous radiation of energy for periods exceeding 3 minutes duration for the purpose of transmission of control signals shall be limited to control functions requiring at least one or more changes during each minute of such transmission. However, while it is acutally being used to control model aircraft in flight by means of interrupted tone modulation of its carrier, a citizens radio station may transmit a continuous carrier without being simultaneously modulated if the presence or absence of the carrier also performs a control function. An exception to the limitations contained in this paragraph may be authorized upon a satisfactory showing that a continuous control signal is required to perform a control function which is necessary to insure the safety of life or property.

§ 95.93 Tests and adjustments.

All tests or adjustments of citizens radio transmitting equipment involving an external connection to the radio frequency output circuit shall be made using a nonradiating dummy antenna. However, a brief test signal, either with or without modulation, as appropriate, may be transmitted when it is necessary to adjust a transmitter to an antenna for a new station installation or for an existing installation involving a change of antenna or change of transmitters, or when necessary for the detection, measurement, and suppression of harmonic or other spurious radiation. Test transmissions using a radiating antenna shall not exceed a total of 1 minute during any 5-minute period, shall not interfere with communications already in progress on the operating frequency, and shall be properly identified as required by § 95.95, but may otherwise by unmodulated as appropriate.

§ 95.95 Station identification.

(a) The call sign of a citizens radio station shall consist of three letters followed by four digits.

(b) Each transmission of the station call sign shall be made in the English language by each unit, shall be complete, and each letter and digit shall be separately and distinctly transmitted. Only standard phonetic alphabets, nationally or internationally recognized, may be used in lieu of pronunciation of letters for voice transmission of call signs. A unit designator or special identification may be used in addition to the station call sign but not as a substitute therefor.

(c) Except as provided in paragraph (d) of this section, all transmissions from each unit of a citizens radio station shall be identified by the transmission of its assigned call sign at the beginning and end of each transmission or series of transmissions directed to or exchanged with a unit of the same station or units of other stations. Each required identification shall include not only the call sign of the station unit transmit-

ting, but also the call sign of the station or stations with which the transmitting unit is communicating, or attempting to communicate. In the case of communications between units of the same station (intrastation), after identifying itself by its assigned call sign, the transmitting unit may identify the other units by unit designators. For communications between units of different stations (interstation), the complete sign of all stations involved must be transmitted. If the call sign of the station being called is not known, the name or trade name may be used, but when contact has been made the called station shall thereafter be identified by its call sign. Examples of proper identification procedure are set forth at the end of this paragraph. Where transmissions or exchanges of transmissions of greater length are permitted by this part, the identification shall also be transmitted at least every 15 minutes. Each transmission or exchange of transmissions conducted on different frequencies shall be fully and separately identified in accordance with the foregoing on each frequency used.

EXAMPLES OF PROPER IDENTIFICATION
Intrastation communications:
(1) Calling: "KZZ 0001 base, calling unit 2."
Response: "KZZ 0001 unit 2, to base, over."
Clearing: "KZZ 0001 base, clear with unit 2" and "KZZ 0001 unit 2, clear with base."
(2) Calling: "KZZ 0001 unit 1, calling unit 3."
Response: "KZZ 0001 unit 3, to unit 1, over."
Clearing: "KZZ 0001 unit 1, clear with unit 3" and "KZZ 0001 unit 3, clear with unit 1."
Interstation communications:
Calling: "KZZ 0001 calling KZZ 0002," or "KZZ 0001 calling KZZ 0002 unit 3" (if appropriate).
Response: "KZZ 0002 to KZZ 0001, over."
Clearing: "KZZ 0001 clear with KZZ 0002," and "KZZ 0002 clear with KZZ 0001."

(d) Unless specifically required by the station authorization, the transmissions of a citizens radio station need not be identified when the station (1) is a Class A station which automatically retransmits the information received by radio from another station which is properly identified or (2) is not being used for telephony emission.

(e) In lieu of complying with the requirements of paragraph (c) of this section, Class A base stations, fixed stations, and mobile units when communicating with base stations may identify as follows:

(1) Base stations and fixed stations of a Class A radio system shall transmit their call signs at the end of each transmission or exchange of

transmissions, or once each 15-minute period of a continuous exchange of communications.

(2) A mobile unit of a Class A station communicating with a base station of a Class A radio system on the same frequency shall transmit once during each exchange of transmissions any unit identifier which is on file in the station records of such base station.

(3) A mobile unit of Class A stations communicating with a base station of a Class A radio system on a different frequency shall transmit its call sign at the end of each transmission or exchange of transmissions, or once each 15-minute period of a continuous exchange of communications.

§ 95.97 Operator license requirements.

(a) No operator license is required for the operation of a citizens radio station except that stations manually transmitting Morse Code shall be operated by the holders of a third or higher class radiotelegraph operator license.

(b) Except as provided in paragraph (c) of this section, all transmitter adjustments or tests while radiating energy during or coincident with the construction, installation, servicing, or maintenance of a radio station in this service, which may affect the proper operation of such stations, shall be made by or under the immediate supervision and responsibility of a person holding a first- or second-class commercial radio operator license, either radiotelephone or radio telegraph, as may be appropriate for the type of emission employed, and such person shall be responsible for the proper functioning of the station equipment at the conclusion of such adjustments or tests. Further, in any case where a transmitter adjustment which may affect the proper operation of the transmitter has been made while not radiating energy by a person not the holder of the required commercial radio operator license or not under the supervision of such licensed operator, other than the factory assembling or repair of equipment, the transmitter shall be checked for compliance with the technical requirements of the rules by a commercial radio operator of the proper grade before it is placed on the air.

(c) Except as provided in § 95.53 and in paragraph (d) of this section, no commercial radio operator license is required to be held by the person performing transmitter adjustments or tests during or coincident with the construction, installation, servicing, or maintenance of Class C transmitters, or Class D transmitters used at stations authorized prior to May 24, 1974: *Provided,* That there is compliance with all of the following conditions:

(1) The transmitting equipment shall be crystal-controlled with a crystal capable of maintaining the station frequency within the prescribed tolerance;

(2) The transmitting equipment either shall have been factory assembled or shall have been provided in kit form by a manufacturer who provided all components together with full and detailed instructions for their assembly by non-factory personnel;

(3) The frequency determining elements of the transmitter, including the crystal(s) and all other components of the crystal oscillator circuit, shall have been preassembled by the manufacturer, pretuned to a specific available frequency, and sealed by the manufacturer so that replacement of any component or any adjustment which might cause off-frequency operation cannot be made without breaking such seal and thereby voiding the certification of the manufacturer required by this paragraph;

(4) The transmitting equipment shall have been so designed that none of the transmitter adjustments or tests normally performed during or coincident with the installation, servicing, or maintenance of the station, or during the normal rendition of the service of the station, or during the final assembly of kits or partially preassembled units, may reasonably be expected to result in off-frequency operation, excessive input power, overmodulation, or excessive harmonics or other spurious emissions; and

(5) The manufacturer of the transmitting equipment or of the kit from which the transmitting equipment is assembled shall have certified in writing to the purchaser of the equipment (and to the Commission upon request) that the equipment has been designed, manufactured, and furnished in accordance with the specifications contained in the foregoing subparagraphs of this paragraph. The manufacturer's certification concerning design and construction features of Class C or Class D station transmitting equipment, as required if the provisions of this paragraph are invoked, may be specific as to a particular unit of transmitting equipment or general as to a group or model of such equipment, and may be in any form adequate to assure the purchaser of the equipment or the Commission that the conditions described in this paragraph have been fulfilled.

(d) Any tests and adjustments necessary to correct any deviation of a transmitter of any Class of station in this service from the technical requirements of the rules in this part shall be made by, or under the immediate supervision of, a person holding a first- or second-class commercial operator license, either radiotelephone or radiotelegraph, as may be appropriate for the type of emission employed.

§ 95.101 Posting station license and transmitter idetification cards or plates.

(a) The current authorization, or a clearly legible photocopy thereof, for each station (including units of a Class C or Class D station) operated at a fixed location shall be posted at a conspicuous place at the principal fixed location from which such station is controlled, and a photocopy of such authorization shall also be posted at all other fixed locations from which the station is controlled. If a photocopy of the authorization is posted at the principal control point, the location of the original shall be stated on that photocopy. In addition, an executed Transmitter Identification Card (FCC Form 452–C) or a plate of metal or other durable substance, legibly indicating the call sign and the licensee's name and address, shall be affixed, readily visible for inspection, to each transmitter operated at a fixed location when such transmitter is not in view of, or is not readily accessible to, the operator of at least one of the locations at which the station authorization or a photocopy thereof is required to be posted.

(b) The current authorization for each station operated as a mobile station shall be retained as a permanent part of the station records, but need not be posted. In addition, an executed Transmitter Identification Card (FCC Form 452–C) or a plate of metal or other durable substance, legibly indicating the call sign and the licensee's name and address, shall be affixed, readily visible for inspection, to each of such transmitters: *Provided,* That, if the transmitter is not in view of the location from which it is controlled, or is not readily accessible for inspection, then such card or plate shall be affixed to the control equipment at the transmitter operating position or posted adjacent thereto.

§ 95.103 Inspection of stations and station records.

All stations and records of stations in the Citizens Radio Service shall be made available for inspection upon the request of an authorized representative of the Commission made to the licensee or to his representative (see § 1.6 of this chapter). Unless otherwise stated in this part, all required station records shall be maintained for a period of at least 1 year.

§ 95.105 Current copy of rules required.

Each licensee in this service shall maintain as a part of his station records a current copy of Part 95, Citizens Radio Service, of this chapter.

§ 95.107 Inspection and maintenance of tower marking and lighting, and associated control equipment.

The licensee of any radio station which has an antenna structure required to be painted and illuminated pursuant to the provisions of section 303(q) of the Communications Act of 1934, as amended, and Part 17 of this chapter, shall perform the inspection and maintain the tower marking and lighting, and associated control equipment, in accordance with the requirements set forth in Part 17 of this chapter.

§ 95.111 Recording of tower light inspections.

When a station in this service has an antenna structure which is required to be illuminated, appropriate entries shall be made in the station records in conformity with the requirements set forth in Part 17 of this chapter.

§ 95.113 Answers to notices of violations.

(a) Any licensee who appears to have violated any provision of the Communications Act or any provision of this chapter shall be served with a written notice calling the facts to his attention and requesting a statement concerning the matter. FCC Form 793 may be used for this purpose.

(b) Within 10 days from receipt of notice or such other period as may be specified, the licensee shall send a written answer, in duplicate, direct to the office of the Commission originating the notice. If an answer cannot be sent nor an acknowledgment made within such period by reason of illness or other unavoidable circumstances, acknowledgment and answer shall be made at the earliest practicable date with a satisfactory explanation of the delay.

(c) The answer to each notice shall be complete in itself and shall not be abbreviated by reference to other communications or answers to other notices. In every instance the answer shall contain a statement of the action taken to correct the condition or omission complained of and to preclude its recurrence. If the notice relates to violations that may be due to the physical or electrical characteristics of transmitting apparatus, the licensee must comply with the provisions of § 95.53, and the answer to the notice shall state fully what steps, if any, have been taken to prevent future violations, and, if any new apparatus is to be installed, the date such apparatus was ordered, the name of the manufacturer, and the promised date of delivery. If the installation of such apparatus requires a construction permit, the file number of the application shall be given, or if a file number has not been assigned by the Commission, such identification shall be given as will permit ready identification of the application. If the notice of violation relates to lack of attention to or improper operation of the transmitter, the name and license number of the operator in charge, if any, shall also be given.

§ 95.115 False signals.

No person shall transmit false or deceptive communications by radio or identify the station he is operating by means of a call sign which has not been assigned to that station.

§ 95.117 Station location.

(a) The specific location of each Class A base station and each Class A fixed station and the specific area of operation of each Class A mobile station shall be indicated in the application for license. An authorization may be granted for the operation of a Class A base station or fixed station in this service at unspecified temporary fixed locations within a specified general area of operation. However, when any unit or units of a base station or fixed station authorized to be operated at temporary locations actually remains or is intended to remain at the same location for a period of over a year, application for separate authorization specifying the fixed location shall be made as soon as possible but not later than 30 days after the expiration of the 1-year period.

(b) A Class A mobile station authorized in this service may be used or operated anywhere in the United States subject to the provisions of paragraph (d) of this section: *Provided,* That when the area of operation is changed for a period exceeding 7 days, the following procedure shall be observed:

(1) When the change of area of operation occurs inside the same Radio District, the Engineer in Charge of the Radio District involved and the Commission's office, Washington, D.C., 20554, shall be notified.

(2) When the station is moved from one Radio District to another, the Engineers in Charge of the two Radio Districts involved and the Commission's office, Washington, D.C., 20554, shall be notified.

(c) A Class C or Class D mobile station may be used or operated anywhere in the United States subject to the provisions of paragraph (d) of this section.

(d) A mobile station authorized in this service may be used or operated on any vessel, aircraft, or vehicle of the United States: *Provided,* That when such vessel, aircraft, or vehicle is outside the territorial limits of the United States, the station, its operation, and its operator shall be subject to the governing provisions of any treaty concerning telecommunications to which the United States is a party, and when within the territorial limits of any foreign country, the station shall be subject also to such laws and regulations of that country as may be applicable.

§ 95.119 Control points, dispatch points, and remote control.

(a) A control point is an operating position which is under the control and supervision of the licensee, at which a person immediately responsible for the proper operation of the transmitter is stationed, and at which adequate means are available to aurally monitor all transmissions and to render the transmitter inoperative. Each Class A base or fixed station shall be provided with a control point, the location of which will be specified in the license. The location of the control point must be the same as the transmitting equipment unless the application includes a request for a different location. Exception to the requirement for a control point may be made by the Commission upon specific request and justification therefor in the case of certain unattended Class A stations employing special emissions pursuant to § 95.47(e). Authority for such exception must be shown on the license.

(b) A dispatch point is any position from which messages may be transmitted under the supervision of the person at a control point who is responsible for the proper operation of the transmitter. No authorization is required to install dispatch points.

(c) Remote control of a Citizens radio station means the control of the transmitting equipment of that station from any place other than the location of the transmitting equipment, except that direct mechanical control or direct electrical control by wired connections of transmitting equipment from some other point on the same premises, craft, or vehicle shall not be considered remote control. A Class A base for fixed station may be authorized to be used or operated by remote control from another fixed location or from mobile units: *Provided,* That adequate means are available to enable the person using or operating the station to render the transmitting equipment inoperative from each remote control position should improper operation occur.

(d) Operation of any Class C or Class D station by remote control is prohibited.

§ 95.121 Civil defense communications.

A licensee of a station authorized under this part may use the licensed radio facilities for the transmission of messages relating to civil defense activities in connection with official tests or drills conducted by, or actual emergencies proclaimed by, the civil defense agency having jurisdiction over the area in which the station is located: *Provided,* That:

(a) The operation of the radio station shall be on a voluntary basis.

(b) [Reserved]

(c) Such communications are conducted under the direction of civil defense authorities.

(d) As soon as possible after the beginning of such use, the licensee shall send notice to the Commission in Washington, D.C., and to the Engineer in Charge of the Radio District in which the station is located, stating the nature of the communications being transmitted and the duration of the special use of the station. In addition, the Engineer in Charge shall be notified as soon as possible of any change in the nature of or termination of such use.

(e) In the event such use is to be a series of pre-planned tests or drills of the same or similar nature which are scheduled in advance for specific times or at certain intervals of time, the licensee may send a single notice to the Commission in Washington, D.C., and to the Engineer in Charge of the Radio District in which the station is located, stating the nature of the communciations to be transmitted, the duration of each such test, and the times scheduled for such use. Notice shall likewise be given in the event of any change in the nature of or termination of any such series of tests.

(f) The Commission may, at any time, order the discontinuance of such special use of the authorized facilities.

SUBPART E—OPERATION OF CITIZENS RADIO STATIONS IN THE UNITED STATES BY CANADIANS

§ 95.131 Basis purposes and scope.

(a) The rules in this subpart are based on, and are applicable solely to the agreement (TIAS #6931) between the United States and Canada, effective July 24, 1970, which permits Canadian stations in the General Radio Service to be operated in the United States.

(b) The purpose of this subpart is to implement the agreement (TIAS #6931) between the United States and Canada by prescribing rules under which a Canadian licensee in the General Radio Service may operate his station in the United States.

§ 95.133 Permit required.

Each Canadian licensee in the General Radio Service desiring to operate his radio station in the United States, under the provisions of the agreement (TIAS #6931), must obtain a permit for such operation from the Federal Communications Commission. A permit for such operation shall be issued only to a person holding a valid license in the General Radio Service issued by the appropriate Canadian governmental authority.

§ 95.135 Application for permit.

(a) Application for a permit shall be made on FCC Form 410-B. Form 410-B may be obtained from the Commission's Washington, D.C., office or from any of the Commission's field offices. A separate application form shall be filed for each station or transmitter desired to be operated in the United States.

(b) The application form shall be completed in full in English and signed by the applicant. The application must be filed by mail or in person with the Federal Communications Commission, Gettysburg, Pa. 17325, U.S.A. To allow sufficient time for processing, the application should be filed at least 60 days before the date on which the applicant desires to commence operation.

(c) The Commission, at its discretion, may require the Canadian licensee to give evidence of his knowledge of the Commission's applicable rules and regulations. Also the Commission may require the applicant to furnish any additional information it deems necessary.

§ 95.137 Issuance of permit.

(a) The Commission may issue a permit under such conditions, restrictions and terms as it deems appropriate.

(b) Normally, a permit will be issued to expire 1 year after issuance but in no event after the expiration of the license issued to the Canadian licensee by his government.

(c) If a change in any of the terms of a permit is desired, an application for modification of the permit is required. If operation beyond the expiration date of a permit is desired an application for renewal of the permit is required. Application for modification or for renewal of a permit shall be filed on FCC Form 410-B.

(d) The Commission, in its discretion, may deny any application for a permit under this subpart. If an application is denied, the applicant will be notified by letter. The applicant may, within 30 days of the mailing of such letter, request the Commission to reconsider its action.

§ 95.139 Modification or cancellation of permit.

At any time the Commission may, in its discretion, modify or cancel any permit issued under this subpart. In this event, the permittee will be notified of the Commission's action by letter mailed to his mailing address in the United States and the permittee shall comply immediately. A permittee may, within 30 days of the mailing of such letter, request the Commission to reconsider its action. The filing of a request for reconsideration shall not stay the effectiveness of that action, but the Commission may stay its action on its own motion.

§ 95.141 Possession of permit.

The current permit issued by the Commission, or a photocopy thereof, must be in the possession of the operator or attached to the transmitter. The license issued to the Canadian licensee by his government must also be in his possession while he is in the United States.

§ 95.143 Knowledge of rules required.

Each Canadian permittee, operating under this subpart, shall have read and understood this Part 95, Citizens Radio Service.

§ 95.145 Operating conditions.

(a) The Canadian licensee may not under any circumstances begin operation until he has received a permit issued by the Commission.

(b) Operation of station by a Canadian licensee under a permit issued by the Commission must comply with all of the following:

(1) The provisions of this subpart and of Subparts A through D of this part.

(2) Any further conditions specified on the permit issued by the Commission.

§ 95.147 Station identification.

The Canadian licensee authorized to operate his radio station in the United States under the provisions of this subpart shall identify his station by the call sign issued by the appropriate authority of the government of Canada followed by the station's geographical location in the United States as nearly as possible by city and state.

FCC FIELD OFFICES

Mailing addresses for Commission Field Offices are listed below. Street addresses can be found in local directories under "United States Government."

FIELD ENGINEERING OFFICES

Address all communications to Engineer in Charge, FCC

Alabama, Mobile 36602
Alaska, Anchorage (P.O. Box 644) 99501
California, Los Angeles 90012
California, San Diego 92101
California, San Francisco 94111
California, San Pedro 90731
Colorado, Denver 80202
District of Columbia, Washington 20554
Florida, Miami 33130
Florida, Tampa 33602
Georgia, Atlanta 30303
Georgia, Savannah (P.O. Box 8004) 31402
Hawaii, Honolulu 96808
Illinois, Chicago 60604
Louisiana, New Orleans 70130

Maryland, Baltimore 21202
Massachusetts, Boston 02109
Michigan, Detroit 48226
Minnesota, St. Paul 55101
Missouri, Kansas City 64106
New York, Buffalo 14203
New York, New York 10014
Oregon, Portland 97204
Pennsylvania, Philadelphia 19106
Puerto Rico, San Juan (P.O. Box 2987) 00903
Texas, Beaumont 77701
Texas, Dallas 75202
Texas, Houston 77002
Virginia, Norfolk 23510
Washington, Seattle 98104

COMMON CARRIER FIELD OFFICES

Address all communications to;
Chief, Common Carrier Field Office, FCC
St. Louis, Missouri 63102
New York, New York 10007

PHONETIC PRONUNCIATION
OF THE ALPHABET

Radio Operators frequently use the following phonetic pronunciation of the alphabet to be more clearly understood.

Letter	Word Substituted
A	Alfa
B	Bravo
C	Charlie
D	Delta
E	Echo
F	Foxtrot
G	Golf
H	Hotel
I	India
J	Juliet
K	Kilo
L	Lima
M	Mike
N	November
O	Oscar
P	Papa
Q	Quebec
R	Romeo
S	Sierra
T	Tango
U	Uniform
V	Victor
W	Whiskey
X	X-ray
Y	Yankee
Z	Zulu

GLOSSARY

CB RELATED TERMS

ANL	Automatic noise limiter
BASE	intended for use in one place
BEAM	Type of highly directional antenna
CB	Citizens Band, the common name of the Citizens Radio Service
CHANNEL	Common name for a CB frequency
COAX	Coaxial cable used to connect the antenna to the transceiver
CRYSTAL	A piece of quartz used to control frequency
DECIBEL (db)	Unit of measure for the loudness of sound
DX	Long distance
FREQUENCY	The pitch of a radio signal that distinguishes it from another
Hz	Hertz (cycles per second)
KHz	Kilohertz (kilocycles) or thousands of cycles per second
LSB	Lower sideband
MHz	Megahertz (megacycles) or millions of cycles per second
MICROVOLT (uV)	One millionth of a volt
MOBILE	Any set intended for use while in motion, as in any vehicle
NOISE BLANKER	See noise limiter
NOISE LIMITER	A circuit that reduces noise from man-made devices
PA	Public address

PEP	Peak envelope power, applies only to SSB transceivers
PL-259	Connector used to connect the coaxial antenna line to the transceiver
RF	Radio frequency signals above 15 KHz
RFI	Radio frequency interference
S-UNIT	Units (from 1–9) indicating the relative strength of a received radio signal.
SKIP	A radio signal, reflected by the ionosphere which is bounced back to earth at a far distant point.
S/N	Signal-to-noise ratio
S+N/N	Signal plus noise-to-noise ratio
SSB	Single sideband
SQUELCH	Circuiting that quiets the speaker until a signal is received
SUPERHET	Superheterodyne circuit, commonly used for its high sensitivity and selectivity
SWR	Short for VSWR; see VSWR
TVI	Television interference
UHF	Ultrahigh frequency; 300–3000MHz
USB	Upper sideband
VAC	Volts alternating current
VDC	Volts direct current
VHF	Very high frequency; 30–300 MHz
VSWR	Voltage standing wave ratio. A rating of the efficiency of an antenna. VSWR of 1:1 is ideal, but rarely achieved. The lower the VSWR the better, as more transmitter power is going into the antenna.

CB JARGON

CB'ers long ago adopted the "10-codes," a system of codes, 10-followed by a number, used to communicate certain common messages, which is still used today by anyone using two-way radio.

Since then, CB radio has become an important part of the on-the-road life of millions of Americans. Most particularly, the long haul truck drivers, have developed their own expressive, funny and logical language of words and phrases, which has become the unofficial language of CB.

"All the good numbers"—Best wishes.

Back door—Last vehicle in a group in communication with each other.

"Back 'em down"—Slow down to legal limit.

Back out—Stop transmitting.

Bear—A minion of the law.

Bear Cave—Police station or barracks.

Bear bite—Traffic ticket.

Bear in the air—Police patrolling in helicopters.

"Beat-the-bushes"—The lead vechicle looks for Smokey to relay his "twenty."

Beaver—Female.

Breaker—Someone who wants to interrupt a conversation.

Bushel—One bushel equals ½ ton; 20 tons is a 40 bushel load.

Camera—Radar unit.

"Catch ya on the flip-flop"—Talk to you on the return trip.

Chase 'em up—The chase car of a 2 car radar set-up.

Chicken coop—Roadside truck weighting station, despised by truckers.

Clean—No police in the immediate area.

Clear—Off the channel; final transmission.

Comeback—Return call.

"C'mon", "come on back"—Invitation to reply.

Comic book—Trucker's logbook.

Copy—Do you understand?

Cotton-picker—Substitute for any expletive (no swearing is allowed on CB).

County-Mounty—Local sheriff or deputy.

Definitely—Emphatically.

"Don't feed the bears"—Try not to pick up any tickets.

Double nickel—55 mile per hour speed limit.

"Down and gone"—Stopped transmitting or moving to another channel.

Dirty side—Eastern seaboard.

Ears—Equipped with CB.

Eatem' up—Restaurant.

18-wheeler—Commonly known as a "semi" or 18 wheeled tractor/trailer.

Eyeball—Visual contact.

Flip-flop, flipper—Return trip.

Foot warmer—Linear amplifier.

Four—10-4, abbreviated; OK?

Four-ten—Emphatic 10-4.

4-wheeler—Passenger car or truck with 4 wheels.

Front door—Lead vehicle in a group in communication with each other.

Good buddy—Universal reference to someone else with a CB.

Grass—Median strip or side of the road.

Green stamp—Fines or a toll road.

Hammer down—Cruising above the speed limit.

Handle—On-the-air nicknames used by CB'ers.

Heater—Illegal linear amplifier used to gain extra range (see also linear).

Hole in the wall—Tunnel.

How 'bout it?—Asking for a response.

Kenosha Cadillac—Any product of American Motors.

Linear—Same as heater.

Land line—Telephone.

Load of Postholes—Empty truck.

Legalizin'—Keeping within the speed limit (opposite of streakin').

Local Yokel—City or town police officer.

Makin' 'the trip—Getting the signal out.

"Mercy!"—Wow; Pause. "Mercy" has no clearly defined meaning.

Modulating—Talking.

Negatory, negative—No.

On the move—In motion.

On the side—Standing by on the channel; listening on the channel.

One time—A short contact.

Over the shoulder—Behind you.

Pedal on the metal—Flat-out; cruising in excess of 55 mph. (see Hammer down).

Plain wrapper—Unmarked police car of (*fill in*) color.

Picture taker—Policeman with radar.

Pick 'em up—Pickup truck.

Portable barnyard—Semi hauling livestock.

Portable parkin' lot—New car carrier.

Pounds—S-units-9S-units on the meter is 9 pounds etc.

Radar Alley—Ohio turnpike.

"Rake the leaves"—Same as back door.

Ratchet-jaw—Non-stop talker.

Reefer—Refrigerated truck.

Rest 'em up—Rest stop.

Rig—CB radio or vehicle.

Rockin' chair—Vehicles between the front and back door.

Roller skate—Small car.

Runnin' barefoot—Legal CB operation.

Sandbox—Dump truck carrying dirt or stones.

Southbounder—Anyone traveling South.

Seatcover—Occupants of a vehicle, usually a female.

Seven-thirds, 73'rds—Best regards.

Shake the trees—Same as front door and beat the bushes.

Shaky side—West coast.

Short-short—A short time.

Stepped on—Overpowered by a stronger transmission.

Smokey Bear—State police.

Spies in the sky and Hounds on the ground—Aircraft working with pursuit cars.

Streakin'—Speedin'.

10–77—Negative contact.

33—10-33 (emergency).

Three's (3's)—Best regards.

3's and 8's—Lots of best regards.

Tijuana Taxi—Full dress (marked) police car.

2-wheeler—Motorcycle.

Twenty—10-20, location; what is your location?

Uncle Charlie—FCC monitoring team.

Walked on—Same as stepped on.

Wall-to-wall—Full scale on the S-meter.
Wall-to-wall bears—A lot of police.
Watergate City—Washington, D.C. (obviously)
"We gone"—Stopped transmitting or leaving the road.
Wrapper—Color of police car.
XYL—Ex young lady, usually a wife.

COMMERCIAL RADIO STATIONS
BROADCASTING ROAD CONDITIONS

A nighttime network of radio announcements warning drivers of weather conditions, detours and other obstacles along major routes.

Asheville	WWNC	570 kHz
Bismarck	KFYR	550 kHz
Butte	KBOW	550 kHz
Chicago	WMAQ	670 kHz
Dallas/Ft. Worth	WBAP	820 kHz
Des Moines	WHO	1040 kHz
Fargo	KFGO	790 kHz
Los Angeles	KLAC	570 kHz
Minneapolis	WDGY	1130 kHz
Nashville	WSM	650 kHz
New Orleans	WWL	870 kHz
Orlando	WHOO	990 kHz
Portland	KWJJ	1080 kHz
Raleigh	WPTF	680 kHz
Richmond	WRVA	1140 kHz
Sacramento	KRAK	1140 kHz
Salt Lake City	KSL	1160 kHz
Spokane	KGA	1510 kHz
Wheeling	WWVA	1170 kHz

10 CODES USED BY SOME CBERS

10- 1	Weak signal
10- 2	Good signal
10- 3	Stop transmitting
10- 4	Affirmative, OK
10- 5	Relay message to _____
10- 6	Busy
10- 7	Out of service
10- 8	In service
10- 9	Repeat transmission
10-10	Negative
10-11	_____ on duty
10-12	Stand by
10-13	Existing conditions
10-14	Message, information
10-15	Message delivered
10-16	Reply to message
10-17	Enroute
10-18	Urgent
10-19	Contact, in contact with
10-20	Location
10-21	Contact _____ by phone
10-22	Disregard
10-23	Arrived at scene
10-24	Assignment completed
10-25	Report to, meet
10-26	Estimated arrival time
10-27	License or permit information
10-28	Ownership information
10-29	Records check
10-30	Danger, caution
10-31	Pick up
10-32	_____ units needed (Give number and type)
10-33	Help me quick
10-34	Time check
10-36	Time check

Cut out on dotted line

NOTES

1. _____   _____
   Station Call Sign   Unit
   TRANSMITTER IDENTIFICATION CARD
   Citizen's Radio Service

   This Card Attests that Authorization has been Received from the F.C.C.
   for installation and or Operation of the Radio Transmitter to which attached. (Fill in items 1 through 6)

   2. Name of Permittee or licensee_____

   _____

   3. Location(s) of transmitter records_____

   _____

   4. Transmitter operating frequencies. CLASS_____ MC.

   5. Current F.C.C. authorization for this transmitter expires_____

   6. Signature_____
                    (Permittee, licensee, or responsible official thereof)
   Equivalent to F.C.C. Form 452-C (Revised)

Cut out on dotted line

NOTES

NOTES

NOTES

NOTES

NOTES